HOW TO FIX A BROKEN PLANET

Do you want to help save human civilisation? If so, this book is for you.

How to Fix a Broken Planet describes the 10 catastrophic risks that menace human civilisation and our planet, and what we can all do to overcome or mitigate them. It explains what must be done globally to avert each mega-threat and what each of us can do in our own lives to help preserve a habitable world. It offers the first truly integrated world plan of action for a more sustainable human society – and fresh hope. A must-read for anyone seeking sound practical advice on what citizens, governments, companies, and community groups can do to safeguard our future.

Julian Cribb AM is an Australian author and science communicator. His career includes appointments as Scientific Editor for *The Australian* newspaper, Director of National Awareness for the Australian Commonwealth Scientific and Industrial Research Organisation (CSIRO), editor of several newspapers, member of numerous scientific boards and advisory panels, and president of national professional bodies for agricultural journalism and science communication. His published work includes over 9,000 articles, 3,000 science media releases, and 10 books. He has received 32 awards for journalism. His previous books include *Earth Detox* (2021), *Food or War* (2019), *Surviving the 21st Century* (2017), and *The Coming Famine* (2010). As a science writer and a grandparent, Julian Cribb is deeply concerned about the existential emergency facing humanity, and his latest books map hopeful pathways out of our predicament.

'... probably the most important book I have read. I predict it will be a world changer. It needs to be translated into every language on Earth and made urgently available to politicians and community opinion leaders everywhere. The author is a distinguished science communicator with many books and publications on the matters covered here. The book brings together in a short and highly readable volume, his conclusions about the survivability of the human species. It offers the reader a sensible and practical path to rescue our human species from early extinction and offers detailed actions for individuals, community groups and governments.'

Bob Douglas AO, Australian National University

'Humanity for the first time in its history is causing a global crash of civilization and even the possibility of its own extinction. Julian Cribb's pathbreaking new book ... looks squarely at the existential threats to our life-support systems and suggests how individuals and governments should take remedial action. I only hope people will read him and act.'

Paul Ehrlich, author of The Annihilation of Nature.

'... a masterpiece that will have a lasting impact on the culture as people seek ways to be effective planetary citizens. This book delivers a worthy prescription for humanity to embrace common purpose. Together, in gender equal partnership, that's our best chance to save ourselves from ourselves'

Geoff Holland, author of The Hydrogen Age.

'... a must read for the future of the human species, and all the other species with which we share this one and only Earth.'

- Dana Hunnes, author of Recipe for Survival: What You Can Do to Live a Healthier and More Environmentally Friendly Life

'Julian Cribb continues to warn all on this planet of what we are doing to it and the risks. As usual, his work is meticulously documented and tells us what can still be carried out – if the will is there at all levels.'

John Kerin AO, former Australian Minister for Primary Industries and Energy

'A blueprint for humanity surviving and thriving in the 21st century. This book provides the escape ramps for avoiding the dangerous problems facing our species.'

Lyle Lewis, Former Endangered Species Biologist, U.S. Department of Interior

'This is a well-written book offering sensible, idealistic prescriptions for humanity, at this moment of truly desperate polycrisis. If we really are *Homo Sapiens*, we'll listen to Cribb ... And yet Cribb does not give up on humanity; far from it. Whether you are interested in rethinking our species or in practical steps to reduce the level of our dire predicament, this book can help.'

Rupert Read, author of Why Climate Breakdown Matters

'... takes you on a chilling journey through the existential threats facing humanity. Essential reading for the 21st Century.'

Will Steffen, Australian National University

How to Fix a Broken Planet

Advice for Surviving the 21st Century

JULIAN CRIBB

Council for the Human Future

CAMBRIDGE
UNIVERSITY PRESS

CAMBRIDGE
UNIVERSITY PRESS

Shaftesbury Road, Cambridge CB2 8EA, United Kingdom

One Liberty Plaza, 20th Floor, New York, NY 10006, USA

477 Williamstown Road, Port Melbourne, VIC 3207, Australia

314–321, 3rd Floor, Plot 3, Splendor Forum, Jasola District Centre, New Delhi – 110025, India

103 Penang Road, #05–06/07, Visioncrest Commercial, Singapore 238467

Cambridge University Press is part of Cambridge University Press & Assessment, a department of the University of Cambridge.

We share the University's mission to contribute to society through the pursuit of education, learning and research at the highest international levels of excellence.

www.cambridge.org
Information on this title: www.cambridge.org/9781009333412
DOI: 10.1017/9781009333436

First published 2023

A Cataloging-in-Publication data record for this book is available from the Library of Congress.

Library of Congress Cataloging-in-Publication Data
Names: Cribb, Julian, author.
Title: How to fix a broken planet : advice for surviving the 21st century / Julian Cribb, Council for the Human Future.
Description: First Edition. | New York : Cambridge University Press, 2023. | Includes bibliographical references and index.
Identifiers: LCCN 2022036371 (print) | LCCN 2022036372 (ebook) |
ISBN 9781009333412 (Paperback) | ISBN 9781009333436 (eBook)
Subjects: LCSH: Nature – Effect of human beings on. | Twenty-first century – Forecasts. | Global environmental change – Forecasting. | Human ecology – Forecasting. | Economic forecasting. | Social prediction. | Human ecology. | Environmental ethics.
Classification: LCC GF75 .C76 2023 (print) | LCC GF75 (ebook) |
DDC 304.2–dc23/eng20221103
LC record available at https://lccn.loc.gov/2022036371
LC ebook record available at https://lccn.loc.gov/2022036372

ISBN 978-1-009-33341-2 Paperback

CONTENTS

PREFACE

How to Fix a Broken Planet is the sixth in a series of science-based books on the growing human existential emergency – and what we can do about it. The intent behind the whole series is to give people the facts they need to take timely, positive action for the sake of human survival at a time fraught with peril for all.

This book has a particular focus on what we humans can and must do if we are to avoid the catastrophic consequences of our own actions, both at a global and at a personal level.

Many people, especially the young, say they feel a sense of hopelessness when they contemplate the sheer scale of the task of fixing global problems as massive as climate change, global poisoning, pandemics, or nuclear war.

The antidote to hopelessness is action. This book is all about action. It is a distillation of the best advice from innumerable scientific and other trustworthy sources on what to do. However, all action must start with an understanding of the nature of the problem it is intending to solve, the wider context of that threat, and its connections to other threats.

To ignore the dangers, to delay action, to argue among ourselves about the right way forward can only magnify the catastrophe we already face. On the other hand, to act wisely, promptly, together, and with foresight can abate and maybe even prevent disaster – and thus save millions, even billions, of lives, including our own and those of our families and descendants.

It is a noble cause, perhaps the noblest in the long history of our species.

Our world has changed. There is no going back to the 'safe' old world of the mid-twentieth century. Hankering for yesteryear and being sentimental about our past lives will not save us. Only action can.

But that action must first be carefully, wisely, and deeply considered. There is no point in adopting measures to fix one disaster if, by so doing, we spawn disasters even greater, or more certain. Each action must first be examined in terms of every other major threat before it is adopted. It must make none of them worse.

The advice gathered in this book attempts to do this. It certainly does not claim to have *all* the answers. New and better ideas and technologies are coming forth constantly. You are encouraged to devise and share your own. However, the book does bring together many practical suggestions from wise minds worldwide for repairing our broken planet, restoring its vital systems, and securing our own future on it. More importantly, it integrates them to address the overarching threat.

Above all, it explains what each of us can practically do in our own lives to help to achieve this. Thus it offers everyone a sense of hope, purpose, engagement, and opportunity.

The problems we face are global in nature. They cannot be fixed by nations or corporations working alone or in small alliances. They cannot be fixed by small groups of well-intentioned people acting, while others sit and watch or try, from selfish motives, to undermine them. These challenges have to be tackled by humanity as a whole, acting together for the first time in our history. They call for the greatest act of single-minded, collaborative action and caring that humans have ever undertaken.

Fixing our 'broken' planet is not going to be easy. But it is achievable if we all act together, with determination and goodwill.

It will be the greatest thing our species has ever done.

Julian Cribb, Canberra

ACKNOWLEDGEMENTS

This book is the fruit of many minds, much science, learning, and human wisdom. In particular, I wish to acknowledge the advice, inspiration, knowledge, encouragement, and dedication to the human future of the following: Emeritus Professor Bob Douglas; Emeritus Professor Paul Ehrlich; Professor John Hewson; Professor Robyn Alders; Major General Michael Jeffery; Emeritus Professor Will Steffen; Emeritus Professor Ian Lowe; Professor Tilman Ruff; Dr Jane O'Sullivan; Professor Ravi Naidu; Professor Toby Walsh; Professor Gerardo Ceballos; Professor Peter Gleick; Dr Soumiya Swaminathan; Dr Laurie Laybourne-Langton; Dr Dana Hunnes; Dr Matt Lloyd; Bill D'Arcy; John Schmidt; Byron and Elissa Vance; Jenny Goldie; Dr Suzette Searle; Geoff Holland; Brad Collis; Michael Brown; Dr Richard Davies; Dr Ta-Yan Leong; Professor John Williams; John Kerin; Alice Ghent; Phillip Adams; Dr Robyn Williams; Patricia Johnson; Sheresta Saini; Pennie Scott; J. Carl Ganter; Stephen Leahy; Steve Morton; Dr Max Whitten; Dr Joanne Daly; Bob Marshall; Dr Grant Griffiths; and Stephan Wellink.

For the power, inspiration, and leadership of their ideas, words, and deeds in restoring our world, I also wish to acknowledge: David Suzuki; Dr Jane Goodall; Dr Sylvia Earle; David Attenborough; Greta Thunberg; Kate Raworth; Johan Rockström; Hans Joachim Schellnhuber; Michael E. Mann; E. O. Wilson; Rupert Read; Antonio Guterres; George Monbiot; Martin Rees; Paulo Magalhaes; John Church; Paul Crutzen; Matthew England; James E. Hansen;

James Lovelock; Peter Raven; Derek Tribe; Terry Hughes; Ove Hoegh-Guldberg; David Lindenmeyer; Peter Kalmus; David Karoly; Kurt Lambeck; Graeme Pearman; Stefan Ramstorf; Owen Toon; Christiana Figueres; Per Pinstrup-Andersen, Leilani Munter; Bill McKibben; Michel André; Barbara Block; Brad Norman; Joseph Cook; Vreni Haussermann; Shafqat Hussain; Laury Cullen; Mark Kendall; Pilai Poonswad; Francesco Sauro; 'Dr Skywater' Murase; Rohan Pethiyagoda; Olivier Nsengimana; Hosam Zowawi; Tomas Diagne; David Irvine-Halliday; Alexis Belonio; Tim Flannery; Megharaj Mallavarapu; Penny Whetton; Vandana Shiva; R. S. Swaminathan; George Rothschild; Graeme Harris; Major General John Hartley; Admiral Chris Barrie; Tim Fischer; His Holiness Pope Francis; Leonardo DiCaprio; Al Gore; E. F. Schumacher; Matthis Wackernagl; HRH Charles, Prince of Wales (now King Charles III); Gro Harlem Brundtland; Amory Lovins; Hugh Possingham; Lester Brown; Ian Dunlop; Stephen Jay Gould; Lindsay Falvey; Brian Walker; Jonica Newby; Phillip Tobias; David Montgomery; Alf Poulos; Mark Stafford-Smith; Karl Kruszelnicki; Mary E. White; Rachel Carson; Naomi Klein; Elizabeth Kolbert; Douglas Adams; Robert Ardrey; Richard Leakey; Mary Leakey; James Howard Kunstler; Naomi Oreskes; David Wallace-Wells; Paul Hawken; Mark Lynas; Jared Diamond; Tim Flannery; Peter D. Ward; Carl Sagan; Katharine Hayhoe; Daniel Wahl; Roger Hallam; Kimberley White; David Hulme; Amitav Ghosh; Laurie Garrett; Mike Berners-Lee; Jacinda Ardern; Sam Mitchell; Richard Heinberg; David Pimentel; Malcolm McIntosh; Dickson Despommier; Suzannah York; Hazel Henderson; Joan Diamond; Ming Hung Wong; Richard Dawkins; Laurens van der Post; Michael Pollan; Bill Bryson; Bill Mollison; Joel Salatin; Frances Moore Lappé; and Colin Charteris.

Finally, I wish to acknowledge the hard work, dedication, and professionalism of the team at Cambridge University

Press, without whose invaluable help this book – my third with the Press – could not have appeared: Dr Matt Lloyd (Publishing Director), Elle Ferns (Content Manager), Laura Simmons, Sarah Lambert (Senior Editorial Assistant), Anne Elliott-Day (copy-editor), Vidya Ashwin (Project Management Executive, Integra Software Services), Chloe Bradley (publicist), and Niamh Courtier (publicist, Australia).

1 EXISTENTIAL EMERGENCY

The Getting of Wisdom

Velvet night enfolds the African savannah. The last light of day vanished half an hour ago; beneath a panoply of starlight filtered by scudding cloud, a boy picks his way home across the veldt, following a familiar track. As he approaches Black Hill, the place where his family takes shelter at dusk each day, the ground becomes rough and uneven. Limestone boulders litter the grassy slopes leading up to a rocky outcrop, groined by eons of rain and wind into a natural fortress of low cliffs, meandering crevices, shallow caves, and shelters – a place even ferocious predators avoid once darkness has fallen.

His attention focused on the uneven footing, the boy fails to detect the deeper shadow in the tree that overhangs this narrow part of the track, beaten by many feet over long ages. Indeed, the tree itself is hardly to be seen – a black silhouette against the fitful starshine. In daylight the tree appears old, rotten, stripped of foliage, devoid of any place to hide, a gaunt object not to be feared. Only on a moonless night like this is it a menace, as the old woman has often warned. But the youth is daring, lithe, and strong. Home is close. The path underfoot skirts the boulders, weaving among the rising outcrops – all other ways are far more difficult, treacherous, and just as risky. He should not have stayed out so long, hunting, to prove his prowess and pride. As he passes beneath the outstretched arm of the tree, a shadow blacker than the darkness overhead launches itself silently, blotting out the dim vault of the sky. The youth knows a moment's panic, total terror, and excruciating agony before his neck is expertly snapped. Teeth like daggers sink into his face and skull and in a series of brutal tugs the limp body is withdrawn silently into the dry grasses, heaved down the hillside to a lair where a hungry brood awaits.

Huddled safe in their rocky hilltop haven the family wait in vain, wait for the return of day, mourning yet another member of their clan to fall to the ruthless serial killer who stalks them in dreams as well as in reality. Not the first, by any means. One of a long, long line of child victims stretching back tens of thousands of years, hundreds of thousands, millions even.

The killing is real. It took place sometime between 1.8 and 1.5 million years ago. The victim was a child, probably a young male from a small family group of pre-humans who regularly made use of the rocky shelters around Swartkrans, in the bushveld not far from Pretoria and Johannesburg in South Africa. We know how it happened because archaeologist Bob Brain, whose team unearthed the grisly forensic evidence, says:

> *Another insight into this came to light at Swartkrans when we found that the back of the skull of a child had two small round holes in its parietal bones. I noticed that the distance between these holes was matched very closely by that of the lower canines of a fossil leopard from the same part of the cave. My interpretation was that the child had been killed by a leopard, probably by the usual neck-bite, and then picked up with the lower canines in the back of the head and the upper ones in the child's face. It was then carried into the lower parts of the cave, and consumed there.*[1]

Brain's sifting of the remains of other prey animals, especially baboons, revealed that leopards had a habit of chewing the long bones but leaving the hard dome of the skull untouched, a grim testimony of ancient slaughter for a modern humanity which has mostly long since forgotten what it is to be the hunted, rather than the hunter.

Yet, around the same time, and in exactly the same place, another equally remarkable event is taking place: people are discovering the use of fire – and of something far, far more important.

The site of Swartkrans offers the first definitive evidence of the control of fire by pre-humans. In a memorable letter to *Nature* in December 1988, Brain and colleagues reported

During recent excavations of hominid-bearing breccias in the Swartkrans cave altered bones were recovered from Member 3 (about 1.0–1.5 Myr BP [million years before the present]) which seemed to have been burnt. We examined the histology and chemistry of these specimens and found that they had been heated to a range of temperatures consistent with that occurring in campfires. The presence of these burnt bones, together with their distribution in the cave, is the earliest direct evidence for use of fire by hominids in the fossil record.[2]

The dating is imprecise, but the rock strata containing burnt bones and other traces of fire is between 1 million and 1.5 million years old. The child's punctured skull was found in a layer at the same site dated at 1.5 million years, or a bit older.

Although there is no direct link between the actual killing and the use of fire, other than the shared location, the inference that fire was first adopted by humans as a defence against predators such as leopards is reasonable and has been widely accepted by archaeologists. It is prob- able that cooking followed soon after, bringing many health and dietary benefits. All animals are afraid of fire, especially the vast wildfires that rage across the world's savannahs, fuelled by grasses cured to tinder in summer heat, ignited by lightning strikes, and fanned by hot, fierce winds. Even when these fires die down, animals avoid the smouldering areas. There must have been a special threat, and a special fear, that drove these pre-humans – creatures with brains not much larger than that of a modern chimpanzee – to beat down their natural instinct to avoid fire at all costs, to gather up the embers, to carry them carefully back to the home site and there conjure the flames forth again.

Ancestral hominids had walked the grasslands of Africa for at least 6 million years before fire came to Swartkrans. They had no doubt fled wildfires many times and seen other animals, leopards included, do the same. To conquer their own fear of fire, and to exploit the leopard's, was a spectacular leap forward into the age of humanity. To do such a thing requires a very special skill: the ability to look into the future, to imagine a possible threat, and to conceive, in the abstract, a way of meeting it. The site of Swartkrans captures both moments in the phenomenal ascendancy of humans.

This unexceptional, low, grassy African hill, with its rocky crown, marks the birthplace of wisdom.[3] It is also the earliest definitive proof of the will of our kind to survive – and the start of the evolution of that remarkable device, human intelligence, through the billions of man-hours we subsequently spent communicating, learning, exchanging ideas, and testing new technologies as we sat around a campfire, safe from predators. It was fire that helped us think together – and fire that made us human.

I first told this tale in *Surviving the 21st Century*, the book that first identified 10 interconnected mega-risks. I retell it here because it is a parable for our times and reveals elements essential to the survival of *Homo sapiens* at the time of our greatest peril from our greatest foe – ourselves. It describes our tendency as a species to take foolish risks; our ability to understand, foresee, and devise solutions to those risks; and the nature of wisdom itself.

Emergency

We humans are facing the greatest emergency of our entire million-year existence.

This is a crisis compounded of 10 catastrophic risks,[4] each of our own making. These threats are deeply interconnected and are now arriving together. However, their collective scale is so vast and their relationships so complex that few yet understand the peril they place us in.

Combined, these risks menace our very civilisation and, potentially, our survival as a species. The chief driver of these perils is the sheer scale of the human enterprise: overpopulation, overconsumption, pollution, inequality, poor choice of technologies, and poor social arrangements.

The danger is exacerbated by widespread delusion about the scale of the challenges we face, fierce competition over dwindling resources, our pride, and our warlike tendencies. This crisis threatens every person on Earth, woman, child, or man – and will do so for the next six generations at least. Its salient elements include:

- Extinction of species and the degradation of the ecosystems which support life on Earth, on a scale we have never previously witnessed.
- Key resources for living, including soil, water, forests, fish, and certain minerals, are becoming scarce – and the risk of conflict over them is escalating. A global water crisis looms. The oceans are heating, acidifying, and losing their oxygen and life. Fish and forests are disappearing.
- Due to technological 'advances', the threat of nuclear conflict is higher than at any time in our history. Nuclear weapons will be used – unless they are abolished.
- We are approaching the point where the Earth's climate may tip out of our control, pitching us into

hothouse conditions which spell disaster for our food supply, health, and civilisation.

- Global poisoning by human chemical emissions is already out of control; it is five times the scale of our climate emissions, kills 13.7 million people a year, cripples 600 million every year, is reducing our intelligence, and is impacting all life on Earth.
- The world food supply teeters on a knife-edge, imperilled by declining resources, particularly soil and water, the decay of the ecosystem in which it exists, and the loss of the stable climate that allowed agriculture to develop in the first place.
- Despite the lessons of Covid-19, we are still unable to identify and prevent future pandemics. Our civilisation is set up to spread them. More are on the way; some may be man-made.
- Artificial intelligence, killer robots, universal surveillance, bio- and nanosciences, and other advanced technologies are unregulated and out of control. Their misuse poses growing risks to society, freedom, health, and the human future.
- The human population is growing at record speed – 1 per cent (almost 80 million people) a year. Megacities, and civilisation as a whole, face increasing risk of failure as we outrun our resources.
- There is widespread delusion, denial, and failure to recognise the reality of our plight. Misinformation is being actively spread by a misguided or malevolent segment of humanity.

Our interconnected social and economic systems are the unifying factor in the breakdown of the Earth's ability to sustain life. Most countries, their economies, and almost all corporations remain committed to growth in consumption, which is speeding us towards disaster. Our current economic model is broken because it now destroys the very things we (and it) need to survive. It must be redesigned or replaced. Fearful of an uncertain future and distrustful of one another, many countries are now seeking deadly recourse to rearmament.

However, there is much that can be done to curb the danger, limit the threats, reduce the number of lives lost to them, and improve human prosperity and well-being. The Council for the Human Future[5] has called on the governments, companies, and communities of the world to develop and implement an urgent plan of action that addresses all these risks and their combined impact – a Plan for Human Survival. Currently, what we have is a chaotic road to avoidable disaster. The group Common Home of Humanity[6] has called for an 'Earth System Treaty', a legal framework that covers all the risks we presently run and seeks to have humans live within a 'safe operating space'.[7]

Survival

The world needs a 'survival revolution' on a scale far larger and more encompassing than the agricultural, industrial, or computer revolutions that have preceded it. Such a revolution, supported by citizens, governments, and businesses worldwide, must address the existential emergency in its totality, meaning it must encompass *all* of the catastrophic risks, not just a cherry-picked handful of them. It must devise cross-cutting solutions that make none of the other risks worse – a major flaw in current thinking. It must take an integrated, systems approach to a complex problem.

The next 10 chapters in this book each deal with one of the threats. The chapters have three parts. The first describes a catastrophic risk and the scientific evidence for it. The second enumerates solutions that must be taken by nations and by humanity as a whole, based on sound scientific advice. And the third describes what individuals can do in their own lives to improve our prospects of survival, also based on expert advice.

Our common existential emergency cannot be left to national governments, which have so far displayed more ability to dither, procrastinate, and pursue selfish, short-term aims than intent to save humanity.[8] In the end, action must be driven by the 8 to 10 billion citizens of planet Earth who are willing to do all that is necessary to save themselves, their children, and their grandchildren. Only through these billions of concerned citizens acting together can politics, commerce, religion, and society be induced to adopt wisdom – and make the changes essential to our survival as a civilisation and as a species.

Together, we must fix our broken planet – before it is too late to do so.

The Solutions

Here, in no special order, are 10 examples of the 'big picture' solutions which humans need to adopt without delay. More detailed explanations are to be found in the chapters that follow.

1. Outlaw all nuclear weapons, eliminate their stockpiles, and safely recycle or bury their materials. (Chapter 4)
2. End all extraction and use of fossil fuels and their by-products – pesticides, plastics, and petrochemicals – by 2030. Replace them with renewable energy and green chemistry. Rewild half the Earth's land area to draw down and lock up carbon. (Chapters 5 and 8)
3. Create a circular global economy in which every resource is recycled and nothing is lost, wasted, or allowed to pollute. (Chapter 3) Dematerialise the economy, substituting ideas for material goods.

Introduce a 'Green New Deal' that reduces inequality, ensures sustainability, triggers investment, and transitions economies into a new model of prosperity and abundance. Introduce an Earth Standard Currency. (Chapter 13)

4. Develop a renewable global food system, consisting of:
   - Regenerative agriculture
   - Climate-proof urban food systems that use recycled water and nutrients
   - Deep-ocean farming of plants, fish, and marine animals.[9] (See also Chapter 8)

5. Return half of the world's current farmlands to forest or wilderness to end the Sixth Extinction and restore the habitability of the biosphere. Create a 'Stewards of the Earth' programme to carry it out. (Chapters 2 and 8)

6. Create a new Human Right Not to Be Poisoned and a 'Clean Up the Earth Alliance'[10] to eliminate all forms of toxic pollution. Introduce a global inventory and universal safety testing of all manufactured chemicals. (Chapter 6)

7. Introduce a world plan to progressively and voluntarily reduce the human population to a sustainable level. (Chapter 9)

8. Prevent future pandemics by ending environmental destruction (Chapter 2), banning dangerous scientific experiments, discouraging global travel, creating early warning systems, and reducing the human population. (Chapters 7 and 9)

9. Give women a greater role in world and community leadership. Unlike men, women tend not to start wars, wreck oceans, fell forests, ruin landscapes, pollute, and poison everyone. They often take thought for the needs of coming generations.

10. Draw up an 'Earth System Treaty' and integrated
     survival plan at the global level that addresses all the
     risks outlined here and their solutions, to be signed by
     all countries and their governments and open to be
     signed by all citizens, bodies, and corporations if they
     so wish.

The challenge facing humanity is vast and daunting – but
its solutions are feasible, heartening, and inspiring. They
offer hope, whereas denial offers only misery and despair.
Furthermore, they offer prosperity, better work, greater
happiness, and increased well-being for all of humankind.

The question is whether humans still have the intelligence and the wisdom to adopt these solutions. Doing so
will help save literally billions of lives that will otherwise
be needlessly sacrificed.

No government on Earth yet has an explicit policy for
human survival, or a constitution which places our survival first and foremost. They are so blinded by the present
they do not see the need.

There is no plan by humanity as a whole to avert the
danger we face and secure our future. It is time we had one.
The rest of this book explains in plain language what such
a plan might look like – and what we can each do in our
own lives to make it happen.

What You Can Do

The first thing is for each of us, as individuals, to understand that our own personal future is at risk – as well
as that of humanity. This book provides hard scientific
evidence for this statement, and there is plenty more
available. Every day we face mounting risks from global
poisoning, an increasingly violent climate, a shakier food
supply, growing scarcity of resources – like water, soil, fish,
and forests that we need for survival – and from the steady

unravelling of Earth systems, such as oceans and climate, that support life on this planet.

The second thing we can all do is not despair. Humans have met and surmounted many life-threatening challenges in the past – and we are *very* good at it. What we are less good at is acting together, with foresight, in plenty of time to curb the danger and lower the death toll. This time we need to act together.

The third thing we can all do is take action in our own lives to rein in our demand for energy and material goods, thus reducing the poisons they emit. By changing our demand from damaging and unsustainable goods to renewable energy, renewable goods, and renewable food, we can send a signal to the world economy that is so powerful no corporation or government can ignore it.

By making the right changes in our own lives, we can help to repair the world we live in and hand it, unbroken, to our children's children.

2 EXTINCTION ... OR SURVIVAL?

The Problem

In the translucent blue waters of the Indian Ocean, scientists stared, amazed, at the coiling shape of the world's longest animal – a wormlike creature they estimated to be 119 metres (390 feet) in length, twice as long as its closest rival, a jellyfish.[1] The discovery of the siphonophore, as recently as 2020, illustrates vividly how much we humans still have to learn about the natural world that surrounds us – and the difficulty we have in fully comprehending what is befalling it.

One brutal fact about life on Earth is that nearly all species, inevitably, become extinct – to be replaced by others which have evolved from them or their relatives. While dinosaurs no longer roam the Earth, we still eat their descendants in the form of fried chicken – and chickens are, today, probably the most numerous creatures on the face of the planet.[2]

In the normal course of events, about 2 mammal species in 10,000 become extinct every century – yet in the past 500 years, since humans cleared the land and dominated and polluted the planet, nearly 1,000 vertebrate animal species have vanished. The reason for concern is that, all around us, species are disappearing at rates up to 1,000 times faster than normal (see Figure 2.1).[3]

Recent estimates derived from the World Conservation Foundation's Red List of endangered species indicate that the rate of extinction is accelerating. Out of 140,000 different species of animals and plants that have been

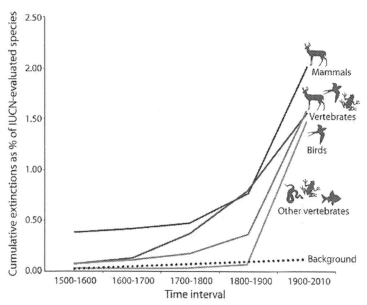

Figure 2.1 Recent extinction rates among mammals, birds, and reptiles are far above normal background rates. They coincide with the growth in human numbers and demand for materials. Source: Ceballos G, Ehrlich PR, Barnosky AD, et al. Accelerated modern human-induced species losses: Entering the sixth mass extinction. Science Advances, 2015, 1 (5).

carefully evaluated by researchers, close to 40,000 met the strict criteria to be classified as endangered. In all likelihood, the unseen extinction of some species – many still unknown to science – has now become a daily event.

Based on this data, scientists conclude that we are now part of the Sixth Extinction event in the Earth's history. In the worst previous event, the Great Dying of the Permian 250 million years ago, 96 per cent of all species died out (see Figure 2.2). Once abundant life forms like trilobites, eurypterids (sea scorpions), and rugose corals have never

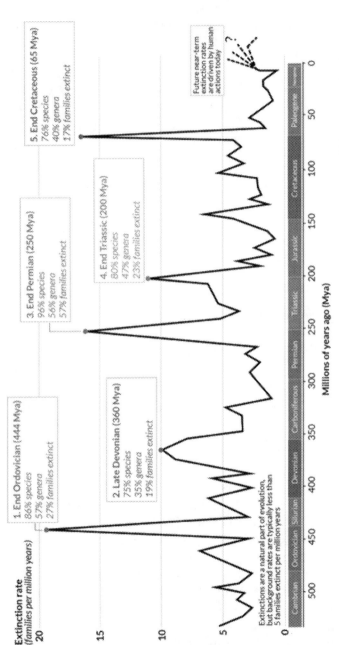

Figure 2.2 There have been five major extinction events in the Earth's history, caused by asteroid impacts, volcanic outbursts, climate change, and global poisoning. Scientists have found evidence we are at the start of the Sixth Extinction event, driven by humans. Source: Ritchie H & Roser M, Our World in Data, 2021.

been seen on the Earth since. Over the grand sweep of time, 99 per cent of all species that have appeared on Earth since life began are gone. Those that survive are chiefly bacteria and algae.

Even among creatures that are not in immediate danger of extinction, populations are crashing worldwide. The World Wildlife Fund (WWF) reported that 69 per cent of all wild animals, birds, and fish perished between 1970 and 2016 as human populations and their material demands exploded across the globe, leading to the clearing and degrading of forests, grasslands, soils, wetlands, rivers, and coastal ecosystems.[4] The rates of loss are not even: they are sharpest in Africa, Asia, and Australasia and are somewhat lower in Europe. In certain ecosystems, such as freshwater rivers and lakes, the loss of water animals is as high as 84 per cent.[5] The destruction has given rise to a new legal term, ecocide – the killing of the systems that sustain life on Earth. It may soon be an internationally recognised crime.[6]

Such wholesale destruction will inevitably rebound on humanity. As the WWF warns: 'Nature is essential for human existence and a good quality of life, providing and sustaining the air, freshwater and soils on which we all depend. It also regulates the climate, provides pollination and pest control and reduces the impact of natural hazards. ... The overexploitation of plants and animals is increasingly eroding nature's ability to provide them in the future.'[7]

Professor Gerardo Ceballos, a biologist at the National University of Mexico, cautions, 'This is perhaps the most urgent and challenging problem that humanity has ever faced'; and, he says, 'Even experts cannot grasp the full scale of the problem.'[8] At current rates of loss, there will be no wild elephants left by 2040. Of all the birds left on Earth, 70 per cent are now farmed poultry. Indeed, of all land animals in our modern world, 96 per cent are now either humans or our livestock. The rest have been wiped out by

our eating habits and the landscapes we have cleared or
turned to desert.

An unexpected outcome of this ecological devastation
is that almost 90 new diseases, such as Covid-19, Ebola, and
HIV/AIDS, have recently crossed from wildlife into
humans – and more are likely to do so.[9] There have been
seven pandemics caused by these invaders since 2000.
Viruses are in a fight for survival too – and if their original
hosts vanish, they have no choice other than to make
people their breeding ground. Figure 2.3 illustrates the

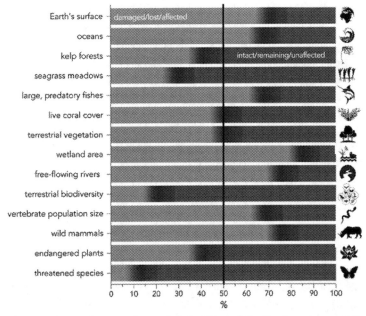

Figure 2.3 How humans have already affected the Earth system and its
ability to support all life. Source: Bradshaw CJA, Ehrlich PR, and Beattie
A et al. Underestimating the challenges of avoiding a ghastly future. *Frontiers
in Conservation Science* 2021, 1, https://doi.org/10.3389/fcosc.2020.615419.
The grey shading shows how much of the key Earth system component is
damaged or lost, and the black how much is intact or unharmed.

profound impact that humans are having on the Earth and its most important life-support systems.

The damage that humans are causing to a habitable Earth is well understood at international policy level, and various measures, such as the United Nations (UN) Convention on Biodiversity,[10] are being developed to tackle it. This Convention calls for significant increases in protected areas to around 30 per cent of the planet's land and sea, while at the same time meeting human needs. And therein lies the problem: the plan offers no solution for continued strong growth in human numbers and demand for materials. It evades the questions of population and growth in consumption, probably for reasons of political expediency. Trying to conserve nature, while consumption and population increasingly command its destruction, is like trying to empty a river with a teacup in a thunderstorm. There is no practical plan to achieve what the Convention espouses and especially for dealing with the most destructive industries – agriculture, forestry, mining, chemicals, and fishing. This is not to derogate the Convention or its aims – rather, it is to say that its goals must be buttressed by action on population, food, resources, and the world economy.

Whatever some people may wish to believe or to hope, we humans are not exempt from extinction. Based on remains discovered to date from the past 4–5 million years, there are some 21 species in the human family tree.[11] And 20 of them are now extinct. As many as eight distinct human species vanished in the last 300,000 years, and three of them – the Neanderthals, Denisovans, and Indonesian 'Hobbit' people (*Homo floresiensis*) – disappeared in just the last 12,000–32,000 years. In all probability, they were wiped out by other humans, either directly by warfare, absorbed by interbreeding, or out-competed for food resources. So, anyone who imagines that *Homo sapiens* is somehow bulletproof and we need not worry about becoming extinct needs to read a little more of our own prehistory.

In *Surviving the 21st Century*, I listed nine credible scenarios for the extinction of our species – eight of them brought about by human action or inaction. They included nuclear war, global overheating to the point where the planet becomes uninhabitable, uncontrollable technology catastrophes, global ecological collapse, global poisoning, mass delusion, disinformation leading to paralysis of the will to solve our problems and survive, and a man-made virus. Optimistically, current humans may be replaced by a wiser species of human, far fewer in number and more rational in its demands on the home planet.[12]

Most likely, it will be a combination of these factors that brings down our civilisation and threatens our survival: famine brought on by climate change and growing shortages of land and water for growing food, vast refugee tsunamis, fresh pandemics, mass poisoning, conventional and nuclear conflict – combining in many different ways. While it is not possible to predict the exact timing or sequence of such events – since they depend on humanity's collective decisions at the time – the drivers are already in motion and the process is under way with the advancing water, climate, ecological, and chemical crises and a renewed arms race.

The initial dominos in the run that leads to civilisational collapse and possible human extinction are already toppling. The grand existential question, the greatest in our entire history as a species, is whether we are wise enough to prevent their fall.

The Solutions

1. *Put back the forests.* Develop a global plan, within an Earth System Treaty, to restore the world's great forests, grasslands, and wetlands; manage the oceans (especially outside

sovereign borders) and cleanse the world's seas, rivers, and freshwater lakes of chemical toxins, plastics, and eroded soil; and restore the chemical balance of the atmosphere for a benign climate.

Pathway: At COP26 in Glasgow in 2021, leaders of 141 nations reaffirmed long-standing commitments to end land clearing and rescue the Earth's forests.[13] But far more is needed than reaffirmation. Action is also essential and that in turn requires investment. Progress in this field has been glacially slow in some areas and has stalled in others. It can be reignited by a global 'Clean Up the Planet' strategy (Chapter 6), by ramping up investment in programmes such as the United Nations Collaborative Programme on Reducing Emissions from Deforestation and Forest Degradation in Developing Countries' (UN-REDD) reafforestation and carbon lock-up scheme,[14] and through a Stewards of the Earth programme (see Chapters 8 and 15).

However, all the things we need to do can be brought together under a single Earth System Treaty, to which all nations would be signatories and which establishes a legal framework for governing our planet by consensus in a safe and sustainable manner. Most homes have their own set of house rules which enable people to live happily together and there is every reason now for the whole planet to develop such a set of rules: the Earth System Treaty would provide the bedrock for a universal code of behaviour which could save and sustain human civilisation while protecting and restoring nature.

2. *Change the world's diet.*

Pathway: Since eating is the most destructive act that humans perform (see Chapter 8), we need to move rapidly to 'renewable food'. This entails replacing at least half the world's farmed and grazed areas with sustainable, climate-proof, intensive food systems located chiefly in cities and in the deep oceans. This will enable the progressive 'rewild-ing' of an area of 25 million square kilometres (similar in size to the continent of North America) and its return to natural forest, grassland, wetland, and wildlife. According to scientific estimates, by designating half of the planet as a nature reserve, we can hope to save 85 per cent of the Earth's animal and plant species.[15] As a bonus, doing this will draw down and lock up a huge amount of climate-destroying carbon, helping to slow down and then progres-sively reverse global heating.

Broadly, this also entails shifting the balance of the world's diet from 'European' (meat and grains) to 'Asian' (vegetables and renewable seafood), which is in any case far healthier. Changing our diet also calls for the rapid shift of half or more of the world's food production capacity:

(a) into cities to protect it from climate shocks and water and nutrient shortages; and

(b) into the deep oceans where large amounts of vegetable and protein seafood can be produced without seriously affecting the rest of the planet.

This demands radical changes in urban planning (to recycle water and nutrient waste into food production); schemes to encourage the uptake of urban food systems at all scales; accelerated investment schemes; and increased R&D into greenhouse, hydroponic, aquaponic, aquaculture, mariculture, and bioculture systems (includ-ing novel food technologies such as synthetic meat, fish, milk, and eggs).

Rewilding also requires a Stewards of the Earth programme, employing indigenous people and former farmers who understand their natural environment and are funded through a global levy on arms and food to repair it[16] (see Chapters 8 and 15). Most plans for rewilding the planet or extending its nature reserves fail to account for the impact of human food production on the environment.

Only by totally redesigning our food systems and making them renewable can we save nature.

3. *Adopt sustainable grazing (also known as 'precision pastoralism') on all of the world's rangelands.* This will enable livestock numbers to be reduced, more carbon to be locked up, vegetation and water cycling restored, and pastoral incomes improved. Sustainable grazing will help to bring back many wild species of plants and animals across the savannahs, which cover 40 per cent of the Earth's land area. It will turn back the spreading of deserts.

Pathway: The concept of 'precision pastoralism' – using satellites and automated mustering to balance feed availability with livestock numbers – allows far more sustainable grazing of rangelands and better incomes for pastoralists. An example of this approach is the sustainable grazing system developed and advocated by Allan Savory.[17] It will need strong support from governments to drive it by ensuring the availability of technology, as well as advice and training for pastoralists and herders across the one-third of the world's land mass consisting of grasslands, prairies, savannah, and rangelands.

4. *Deep ocean farming.* Replace destructive wild-catch fishing and coastal aquaculture with sustainable deep-ocean aquaculture, based on farmed algae and recycled nutrients.

Pathway: Aquaculture is already the world's fastest growing food industry, as wild ocean fish catches dwindle.[18] It will be greatly accelerated by the farming of algae (both seaweeds and single-cell algae) as a major new feed supply for both farmed fish and other livestock as well as for human food, green chemicals, biotextiles, bioplastics, and renewable transport fuel. Large-scale aquaculture in the deep oceans avoids the problems of nutrient pollution, overcrowding, and disease that limit coastal aquaculture. Because it occupies three dimensions, a vastly larger amount of food can be produced in a cubic kilometre of ocean than on a square kilometre of land, greatly reducing the present human footprint on the planet. As a step towards this, decommissioned oil platforms are already being examined as potential bases for deep-water aquaculture.[19] The development of deep-ocean aquaculture will not only avoid coastal pollution, it will also enable the recovery of wild fish populations, which have declined by two-thirds in the sea and 84 per cent in freshwater over the last 50 years.[20]

5. *Replace fossil fuels and petrochemicals with renewable energy, bioplastics, and green chemistry.*

This will eliminate the world's main source of toxic pollution, which causes colossal unseen damage to the health of wildlife and humans – directly and indirectly, through brain poisoning, reproductive dysfunction, developmental disorders, and immune system breakdown – as well as causing climate change.

Pathway: Detailed pathways and options for climate change mitigation have been laid down in many reports by the UN, governments, and informed individuals.[21] They include strategies such as accelerated investment in renewables; carbon cap-and-trade schemes; distributed energy generation; improved energy storage; energy efficiencies in industry; decarbonising of transport and agriculture; sustainable cities and homes; smart energy technologies; electric vehicles; green hydrogen; reafforestation and revegetation of landscapes; and recycling of materials – most of which hold additional benefits for the natural world in terms of reduced toxicity and increased wilderness. Care and forethought must be exercised to ensure that new industries (such as lithium mined for batteries) do not create similar ecological havoc to coal or oil extraction (see Chapters 5 and 6).

6. *Strengthen global biosecurity.* Upgrade the existing global network to identify and combat the introduction and

spread of invasive species, including pandemic human, animal, and plant diseases, in a timelier fashion.

Pathway: Since the wake-up call of Covid-19, moves to improve pandemic detection and forecasting have improved significantly, as have drug and vaccine development. Little, however, has been done to *prevent* new diseases crossing into humans. To properly control the origin and spread of novel plagues, the following measures are required:

(a) an end to land clearing and deforestation globally;
(b) a reduction in world travel by air and sea accompanied by much stricter health rules;
(c) far stronger biosecurity in the import/export of goods;
(d) better control over the transport of marine pests, exotic insects, fungi, and disease-causing organisms;
(e) bans and much stronger public oversight of dangerous scientific experiments such as 'gain of function'[22]; and
(f) bans on the trade and consumption of wildlife to prevent the introduction of exotic diseases to humans (see Chapter 7).

7. *Invest in natural capital.* Build into all food and consumer goods a small charge to fund the repair or prevention of the ecological damage caused by their production. This must be regarded as a wise reinvestment in natural capital, not as an 'eco-tax'.

Pathway: The simplest way to do this is through a small levy on consumer goods, including food, at point of sale, earmarked specifically for reinvestment in natural capital and repair of damaged landscapes and waters. This is equivalent to a VAT or consumption tax, with the salient difference being that it will be used to restore and rebuild the Earth's ability to support us, instead of being consumed by often wasteful government spending. To avoid regressivity, people on low incomes can be exempted or supplied with 'food stamps', subsidies, or other forms of support.

The virtue of such a levy is that those who consume the most make the greatest contribution to renewal. Part of the funds thus raised can be used to pay the world's 1.8 billion farmers and indigenous people to act as Stewards of the Earth (see Chapter 8) and fund conservation, recovery, and rewilding programmes for restoring vital habitats, forests, and key species.

What You Can Do

- Be an informed consumer. Learn which foods[23] and consumer goods degrade and destroy the natural world – and which help to heal it. Exercise your economic power and personal freedom of choice every day to send a clear signal to shops, industry, your nation, and the world economy that you want a clean, safe, and healthy Earth. Freedom is not just a right – it is a responsibility. Every time you spend money you cast your vote on the future of the planet where you and your children will live. It deserves to be cast wisely and with forethought.
- Use the internet and social media to learn the scientific facts about extinction and share them with friends, family, and followers.[24] Play your part as an educator and leader in the online global conservation movement. Stand up for endangered and key species.
- Teach your children about the value of wildlife and natural landscapes and how they support us – and what we lose when we degrade or destroy them. Hand to them a tradition of stewardship – not one of exploitation.
- Support international organisations dedicated to ending the destruction of the world's wildlife and restoring damaged regions, such as the World Wide Fund for Nature (WWF),[25] E. O. Wilson's Half Earth Project,[26] Sylvia Earle's Mission Blue,[27] the Jane

Goodall Institute,[28] Greenpeace,[29] Friends of the Earth,[30] and national bodies such as The Sierra Club.[31]

- Support an Earth System Treaty to repair and regenerate the Earth.
- Support politicians and companies who show a clear track record of devoting real resources to the protection and restoration of wildlife and landscapes. Become sensitised to 'greenwash' (false or exaggerated claims about sustainability made from dubious motives).[32]
- Avoid products that use plastics, pesticides, endocrine disruptors, volatile organic compounds (VOCs), and other poisons that kill, incapacitate, or sterilise wild birds, animals, and fish.
- Choose foods and consumer goods that reduce human pressure on the natural environment and encourage 'rewilding'. For example, choose fresh fruits and vegetables that use few or no pesticides and preservatives. Choose locally grown foods, including urban-produced food, so your preferences can be transmitted direct to producers, rather than muffled by long food chains. Choose to eat more vegetables and less meat.[33] For advice on healthy, sustainable meals, see Dana Hunnes' *Recipe for Survival.*[34]
- Work through local volunteer, social, religious, and sporting groups to repair your local environment – plant trees, restore native species, and share an understanding about how to live more sustainably. The WWF has excellent suggestions on sustainable living.[35]
- Avoid buying stuffed toys. Nowadays, they are mostly made from petroleum and are a form of pollution, soft and cuddly though they seem. Spend the same amount on a good conservation body or activity (like tree planting) and help save or care for a real animal to delight your grandchildren. Involve your children in wildlife sponsorship schemes.

- One group of researchers has identified 27 behaviours which people can easily adopt to reduce their pressure on the environment.[36] These include actions such as:
  - Choose a green energy supplier for your home (and, if possible, your work).
  - Choose to eat local, organically grown fruits, vegetables, and fresh produce.
  - Eat less red meat, and more vegetables, fruit, and grains.
  - Plant and maintain a food garden.
  - Plant and maintain a garden that feeds wildlife as well.
  - Eat and enjoy seasonal produce (as opposed to that which has been stored, frozen for months, or transported round the world at huge environmental cost).
  - Discuss safe, healthy, sustainable eating with your family and circle of contacts.
  - Discuss ways to preserve nature, trees, forests, and wildlife with your family and circle of contacts, so as to regreen cities and rural landscapes.
  - Use recycled paper, timber, textiles, etc.
  - Take part in 'citizen science' projects which study nature and its condition.
  - Avoid using pesticides and poisons at home and at work.
  - Choose your political representatives based on their understanding and willingness to act to save nature.
  - Donate to organisations that work to save endangered species and restore habitats.
  - Be a responsible pet owner. For example, don't allow your dogs or cats to roam unsupervised and destroy wildlife.
  - Don't shoot, hunt, or kill wildlife. At our current population and extinction levels, hunting can be

lethal for many species. Don't buy products such as fur, ivory, bone, or medicines that are made from wild animals.

○ Be an ethical and responsible investor by choosing investments that repair and restore biodiversity and which conserve and replant forests and grasslands.

○ Spend more time in nature with your children, learning and understanding how it protects, sustains, enriches, and supports you.

All of these actions are easy to perform, enjoyable, and rewarding in many ways. They offer a sense of purpose and hope which can overcome the feelings of helplessness and despair that sometimes seize people when they contemplate the awesome scale of the challenges we face. They enable us to prove to ourselves, every day, that we are doing something positive to build a brighter, safer, healthier, happier future for all – and helping to secure our children's place in a habitable world. The more of these actions we do, the better our prospects, the safer our planet.

Never give up. In the words of British environmental writer George Monbiot: 'Almost every day I'm asked "But what can I do?" The answer is the same as it has always been. Combine with other people. Mobilise. Build the movement until it becomes big enough to reach a social tipping point. The key to resistance is persistence.'[37]

3 RESOURCES FOR LIVING

The Problem

Humanity's hunger for resources devours over 100 billion tonnes of materials a year, 12 tonnes for each of us – while people living in well-off countries consume 6 to 10 times as many resources as those in poor ones.[1] Yet very few people are aware how much stuff it takes to support them, or its true cost to the planet. In recent decades, human demand for material goods has exploded – each of us now consumes 10 times what our own great-grandparents needed to live their thrifty lives a century ago.

This is far more than the Earth can support in the long run. Over a lifetime, each of us now:
- uses 35,000 tonnes of fresh water (mostly in the form of food);
- causes the loss of 650 tonnes of topsoil (mainly through food demand);
- uses 120 tonnes of pure energy (oil equivalent);
- wastes 13.5 tonnes of food;
- causes the emission of 119 tonnes of often toxic chemicals;
- causes the emission of 350 tonnes of climate-wrecking CO_2.[2]

This is a colossal personal impact which, collectively, is now running way beyond our planet's capacity to supply, absorb, cleanse, and renew, as many scientists have warned. The Global Footprint Network calculates that we are consuming materials equivalent to the output of

1.75 planet Earths each year – and that we overshoot the globe's renewable resource budget by July each year.[3] The rest of the year is spent running up a deficit that, sooner or later, will be paid for in famine, scarcity, war, disease, and human and animal suffering. Human consumption of material resources is skyrocketing: it has tripled from 29 billion tonnes a year in 1972 to 101 billion tonnes in 2021 – and is on track to reach 170 billion tonnes by 2050.[4]

Resources are the mainspring of unprecedented growth in the old world economy and today's high material living standards. Their growing scarcity is also one of the triggers for civilisational collapse. Somehow, we have to reduce our dependency on them.

Water

The world water crisis is already upon us. Two out of every three humans face acute water scarcity for one month or more a year[5] – the actual number is forecast to reach 5.7 billion by mid-century. Globally, rivers, lakes, and wetlands are dying and drying up, mountain glaciers are shrinking and groundwater reserves are running out in many countries and entire regions, while the human population and its insatiable demand for water continues to rise relentlessly, exacerbated by a warming climate which accelerates the hydrological cycle.

All told, humans now use about 4 trillion cubic metres of water a year. In the time that our population has tripled, our use of water has grown fourfold (see Figure 3.1). This means that the average person now consumes (and wastes) about 500 tonnes of water annually, around two-thirds of which is used to produce our food while much of the rest goes to make the material goods we buy and use – from cement to furniture and clothing. Our personal daily use for washing and drinking is relatively small and is

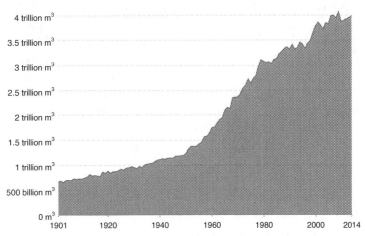

Figure 3.1 Global consumption of fresh water. Source: Our World in Data, Creative Commons, 2021. https://ourworldindata.org/water-use-stress.

therefore not a reliable guide to our actual water impact on the planet; we must look mainly to our buying habits to reduce the water we consume. To put our water use in perspective: you will personally consume enough fresh water in your lifetime to float a battleship.

The amount of fresh water on Earth is finite and has been since the planet was formed. The current scarcity is due mainly to greed, universally bad management, and gross pollution of the resource in almost all societies. The obvious answer – recycling – is only being adopted, marginally, by states such as Singapore and a handful of cities. Water shortages are everybody's problem: while supplies remain adequate in cooler parts of the world such as Europe and North America, even these will be impacted if dangerous water shortages occur in Asia, the Middle East, and Africa – and billions of people are forced to flee their home countries. Water shortages also spell food scarcity and higher world food prices for all.

The global water crisis is the most immediate of the catastrophic risks facing humanity. It is here now, and will only get worse as the century advances without drastic reform, major changes in food production and urban planning, and worldwide personal action to save and reuse precious water.

Forests

The world's forests are dwindling at a rate of 6.6 million hectares a year, and deserts are spreading across 12 million hectares of fertile farmland, every year. The UN Food and Agriculture Organization says:

> *Deforestation and forest degradation continue to take place at alarming rates, which contributes significantly to the ongoing loss of biodiversity. Since 1990, it is estimated that 420 million hectares of forest have been lost through conversion to other land uses, although the rate of deforestation has decreased over the past three decades. Between 2015 and 2020, the rate of deforestation was estimated at 10 million hectares per year, down from 16 million hectares per year in the 1990s. The area of primary forest worldwide has decreased by over 80 million hectares since 1990.*[6]

Forest loss is far more serious for humans that the mere absence of trees. The clearing and thinning of forests for wood and paper production is making global heating worse, as depleted forests change from being carbon absorbers to carbon emitters. It is exacerbating the loss of wild animals, the extinction of many species, and the collapse of the very ecosystems that support humanity. It is destroying rivers, lakes, and wetlands through soil erosion and turbid drinking water supplies. The clearance of large areas of tropical rainforest in the Amazon and Congo basins and south-east Asia is expected to cause significant local and global climatic disruption, including the spread of deserts (see Chapter 5). The drying of cleared landscapes can shatter local farming and food production. As forests dwindle, they also release new diseases into the human population, some of which become pandemic (see Chapter 7).

Soils

The crisis in the world's soils is as great as the crisis in water – and is expanding at similar rates as more land is cleared to feed a burgeoning human population and 'profligate consumerism' grows. Recent estimates indicate that 40 per cent of the planet's land area is degraded.[7] Various scientific studies put the global rate of soil loss at between 36 and 75 billion tonnes a year[8] – meaning that an astonishing 5–10 kilograms of soil are lost for every meal you eat. It also represents the loss of around 1 per cent of the world's farmed soils every year. Sheffield University researchers estimate that a third of our soils are already gone[9] – and over the next half-century, at current rates of destruction, half the world's remaining farmland is going to disappear. This will affect the health and survival of all humans and most of the Earth's living species.

Humanity still relies on agriculture to produce about 95 per cent of our food, but by 2070 we will have only one-third of the farmland we had in 1975 – to furnish twice the amount of food to support a projected 10–11 billion people. As anyone can see, that places our future food security in extreme jeopardy. A system that destroys itself cannot endure. Even more serious than the shortage of water and the increasingly erratic climate, the loss of topsoil combined with the spread of deserts is a gravely underestimated threat to our human future – and, so far, is proving extremely difficult to reverse, or even to slow.[10] However, the UN says that a worldwide effort to restore our vanishing soils is urgent and must be achieved by 2030.[11]

Oceans

In the oceans, 762 polluted 'dead zones' embrace an area of 245,000 square kilometres,[12] and 94 per cent of the world's fisheries are either maxed out or overfished.[13] Acidification (caused by the burning of fossil fuels) threatens the entire oceanic food chain, from plankton and corals up to large sea life.[14] Furthermore, there are alarming signs that the oceans are giving up their oxygen[15] – in other words, their ability to support life – as well as losing their capacity to absorb humanity's excessive carbon emissions. Humans are also dumping around 14 million tonnes of plastic in the oceans every year, along with much other toxic waste, creating a further source of poisoning in the marine food chains that supply our food.[16]

Metals

The world currently mines about 10 billion tonnes of metal ores a year, whose 100–150 billion tonnes of waste cause widespread destruction of landscapes, pollution of water, and harm to the health of humans and wildlife. By mid-century the world's reserves of cheap phosphorus and potash, critical to the global food supply, will run low, endangering food security. While supplies of base metals are deemed adequate to meet future demand, shortages are already emerging among 35 rare earths and metals of strategic value to the electronics and renewable energy sectors, such as lithium.[17]

Overuse

Resource overuse endangers humanity at several levels – through contamination and poisoning of people and wild-life, and through destruction of the natural systems that support us, especially air, water, soil, and biodiversity. Throughout history, resource scarcity and population pressure have often led to war (as was the case with World War II) – indeed, national borders are usually drawn around food-producing resources.

The good news is that we have solutions to all these problems. The challenge is to implement them at global and national scales – as well as by individuals in their own lives. The International Resources Panel of the United Nations Environment Programme (UNEP) has urged the world to decouple the global economy from materials[18] – which means transitioning the economy to rely for its growth on 'things of the mind' rather than the production of excess material goods. This is also known as the 'know-ledge economy'[19] or 'ideas economy', and encompasses fields such as the arts, sciences, sport, entertainment, publication, IT and software, and the teaching and caring professions instead of 'old industries' like mining, manu-facturing, and construction.

Since the 1970s gifted thinkers such as Kate Raworth, Walter Stahel, and Genevieve Reday have been advocating the adoption of a 'circular economy', in which all materials are reused over and over again, and nothing is wasted or left to pollute.[20] By the 2060s it is possible that the human population, if led by the world's women, will begin to contract. If individual consumption can also be reined in, this means that, for the first time in history, our demand for raw materials will also begin to shrink. As recycling takes hold in society, mining, forestry, and the extraction of new materials from nature will become unnecessary: miners and manufacturers will convert themselves into waste harvesters, recyclers, and reprocessors, as some are already doing.[21]

Important concepts in the circular economy include 'cradle-to-cradle manufacturing',[22] zero waste,[23] renewable food production,[24] renewable energy,[25] renewable cities, water and nutrient recycling, and green chemistry. The circular economy is totally achievable. It is profitable, creates new jobs and enterprises and bequeaths a safer environment for us all to live in. It is the way of the future. All we have to do is support the change and recycle everything – as consumers, as investors, and as voters.

The Solutions

1. *Build the 'circular economy'.* Recycle everything that is scarce, especially water, metals, glass, timber, food and organic waste, old clothing, and building materials.

Pathway: Many people now advocate a 'steady state' economy[26] or even a 'degrowth economy'[27] and there is much to commend both ideas. They embrace the idea of a stabilised population (zero growth) and reduced consumption, based on the insight that human happiness has little or nothing to do with production or material 'wealth'.[28] Both stress the need for the human economy to not exceed Earth's ecological limits. However, global consumption is already far beyond those limits (about 70 per cent[29]) implying that there must be a sharp reduction in both human numbers and consumption levels to achieve a steady state. Furthermore, there is mounting evidence that: (a) 'growth' is no longer a viable option due to global resource scarcity; (b) in developed economies growth detracts more than it adds; and (c) growth no longer meets the social goals of employment, equity, health, and habitability.[30]

Notwithstanding this, opposition to a degrowth economy from those invested in the old, material economy is likely to be stubborn – more so, even, that the opposition of the fossil fuels lobby to safe climate measures. However, proponents of a 'steady state' point out that the alternative is far worse – an uncontrolled crash in human numbers caused mainly by famine, disease, mass migration, and war. All this makes the circular or doughnut economy

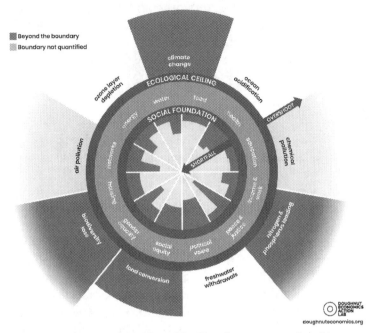

Figure 3.2 Kate Raworth's idea of the 'doughnut economy' enables us to keep human demands within safe planetary boundaries while sustaining and being fair to all. Credit: Kate Raworth and Marcia Mihotich, CC-BY-SA 4.0

approach more pragmatic and acceptable as an immediate way to start reining in our impact on the planet (see Figure 3.2).

The key to a circular economy lies in improving the economics and social attractiveness of recycling through a combination of enhanced consumer demand for recycled products and government incentives. This can be achieved through public awareness and education; by industry marketing its successes; through research and development; and by government incentives, subsidies, and tax breaks. Approaches such as 'industrial ecology' and 'cradle-to-cradle manufacturing' should be widely encouraged and adopted.

2. *Decouple economic growth from the limitations of material resources.* Focus on growth within the 'knowledge economy' and in the products and occupations of the inexhaustible human mind and imagination.

Pathway: The best way to reduce human consumption of material goods is to swing the economy towards making goods and services that are products of the human mind, rather than those made from extracted materials. This should be promoted by government and industry as a 'win–win' for the twenty-first-century economy, which can achieve continued economic and jobs growth while reducing our physical demands on the planet. New material goods can still be produced, but solely by recycling used goods. People with jobs in the 'old' extractive industries can easily retrain for better jobs in the ideas and circular economy.

3. *Price natural resources at their true value and eliminate subsidies.*

Pathway: Governments must be pressured by citizens to eliminate inherent contradictions in their policies, such as subsidising fossil fuels, agriculture, and other resources like water (which encourages wasteful use, increases pollution and sickness, and destroys climate and the environment faster), while at the same time trying to mitigate climate change, water waste, soil loss, and other undesirable impacts.

4. *Phase out the permanent dumping of waste and replace it with 'zero waste' and waste-stream mining.*

Pathway: We need to do away with landfills and rubbish dumps and ban their use. Economic signals that encourage this are essential. Heavier charges, prohibitions, and penalties may be needed to discourage waste production, disposal, dumping, and pollution. Monetary and social incentives are also needed to encourage recycling/reuse, as is widespread consumer and child education. Every enterprise should have a mandatory waste recycling plan.

5. *Progressively replace fossil fuels with renewable, non-polluting energy – such as solar, wind, wave, tide, geothermal, algal oil, thorium cycle nuclear, green hydrogen, and green chemistry.* Eliminate the use of toxic petrochemicals.

Pathway: The market is the best tool to achieve this transition rapidly and determine which of the alternatives are most affordable. However, carbon cap-and-trade, consumer education, incentives leading to greater demand for clean energy, disinvestment in fossil fuel companies (reward them for switching to renewables), elimination of fossil fuel subsidies, tougher environmental restrictions on mining and extraction, and stronger pollution penalties are also necessary.

6. *Adopt a fresh approach to water use that involves rethinking supply, demand, and delivery efficiency.* Develop new ways to provide the goods and services we want using much less water.

Water is both an economic good and a human right. The quality of source water must be protected and matched to the quality we need, and this involves protecting the health of the ecosystem from which the water originates. There should be more community participation and flexibility in water management. There must be a major move towards recycling of all urban water, including wastewater, treated water, and stormwater, moving cities as close to self-sufficiency as possible.

There are huge opportunities to improve water-use efficiency throughout society, across industry, and in agriculture. We must restore the natural flow of rivers, to bring back their health and the quality of the water they provide – including ending all dumping of chemicals, pharmaceuticals, and unprocessed human and animal waste. Water institutions must change from managing concrete infra-structure to managing the whole resource, for equity and well as sustainability – not for money or exclusivity. As water expert Professor Peter Gleick says, 'We need new thinking, better technologies, better economics, smarter institutions, effective outreach and partnerships.'[31]

7. *To save the world's soils from inevitable destruction we must replace most of our present large-scale industrial agriculture and grazing.* In its place we need: (a) regenerative farming; (b) intensive, sustainable, climate-proof urban food production systems; and (c) aquaculture and algae culture in the deep oceans.
    Pathway: See Chapter 8 for more detail.

8. *Strengthen worldwide cooperation on the governance, protection, and restoration of fresh waters, forests, soils, landscapes, and oceans, especially outside national jurisdictions.*
    Pathway: Greater national and local collaboration and knowledge sharing via the internet, UN agencies, international agreements, and farmer-to-farmer.

9. *Provide women with the resources, freedom, and power to plan their families, curb human population growth, and raise and educate fewer children, but better.* Smaller families will lead to reduced demand for material goods, stronger economic growth, and less pollution and ill health (see Chapter 9).
    Pathway: Universal education, healthcare, gender equity, female empowerment, and family planning. This can be delivered collaboratively at global, national, and local levels, also via the internet and social media and by UN agencies, governments, and non-government agencies, and by responsible global corporates.

10. *Place a charge on all resource use, including food, and reinvest it in replanting and regenerating forests, grasslands, desert margins, rangelands, farms, fisheries, rivers, lakes, wetlands, and coastal ecosystems.*

Pathway: This is best achieved via the taxation system or by a method similar to carbon trading, which places a higher cost on less efficient and unsustainable production systems and rewards those that are more efficient and that recycle, regenerate, and renew, thereby encouraging transition. This should be thought of as a wise investment in the future of the planet and its ability to sustain humanity, not as a tax – a reinvestment in the Earth's natural capital. It is essential that farmers receive price signals which facilitate the switch to regenerative agriculture, and miners likewise to metal recycling or renewable forms of energy.

11. *Establish universal, free, education to ensure that every Earth citizen understands the need to sustain and care for the vital Earth systems and natural resources that support them.*

Pathway: Already among the UN Sustainable Development Goals,[32] it is noted that global awareness is needed that education is not just desirable – it will also play a determining role in our human destiny. Education is not a luxury, but a precondition for our surviving the twenty-first century. Countries with free, universal education advance more rapidly. If education is about preparing children for life, then future education systems must include instruction in how to avoid waste and pollution, how to shop more wisely, how to use our economic power to protect and restore the natural world, and how to grow food renewably. Every child should learn hands-on how to produce their own food and how to care for the environment and for animals.

What You Can Do

- Conscientiously reduce your personal and household consumption of material resources, use of energy, and production of waste. Try to keep track of what you buy and what you throw away. Minimise both.

- Take an active interest in how much your lifestyle impacts the environment that supports you using one of the many calculators available, such as that produced by the Global Footprint Network.[33]
- Recycle as much of the waste from your own home as possible through a public recycling system (if available), through recycling companies, and through activities such as home composting. This applies especially to kitchen scraps, old containers and packaging, plastic, bottles, glass and metal cans, paper and cardboard, old clothing, electronic consumer goods and devices, and wooden products.
- Learn about the 'five Rs' – refuse, reduce, reuse, recycle, and rot.[34]
- Learn what goes into the food and goods you purchase and how sustainable they are, so as to become a wiser, less wasteful consumer. Send a signal to regenerative farmers that you value their efforts to produce clean, healthy, sustainable food.[35] Avoid big supermarket chains who sell national and global brands of industrialised food, high in chemicals, wrapped in plastic, produced by unsustainable methods, and shipped needlessly around the world (see Figure 3.3).
- Try to eliminate plastic in all forms from your home and life. It is made from petroleum and it breaks down into micro-fragments which can end up in your own brain, blood supply, and organs, where they cause disease and poisoning as well as harming wildlife across the planet and adding to global heating.
- Choose a diet, activities, and consumer goods that sustain rather than destroy or waste natural resources, and which heal you and your family rather than cause sickness.

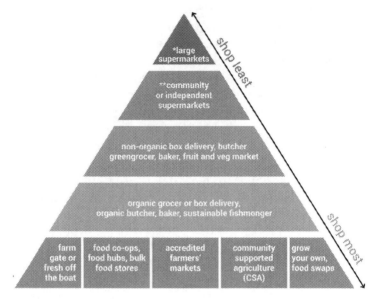

Figure 3.3 Sustainable food shopping guide. Credit: Sustainable Table, 2021

- Participate actively as a global citizen in the universal online consumer and lifestyle movement to share knowledge, ideas, and advice for a sustainable future.
- Use your power as a consumer to send a signal to manufacturers, farmers, industry, and government that we value sustainable goods and services – and will reward those who deliver them.
- Use your power as a voter to support politicians whose deeds (not just words) demonstrate they are committed to a sustainable knowledge economy.
- Only vote for politicians who are sincere about ending waste and building a renewable economy and society.
- Live more like your grandparents: consume far less, reduce possessions, repair and keep things longer,

    recycle, waste nothing, rediscover the virtue (and
    satisfaction) of thrift.
- Teach your children to prize clean air, water, soil,
  forests, wildlife, and the ocean as much as life and
  liberty. For that is what they are.

4   NUCLEAR AWAKENING

The Problem

The peril of annihilation has lain like a vast shroud over humanity since the 1950s, with the proliferation of nuclear weapons and the machinery to deliver them practically anywhere on Earth.

The urge by humans to self-destruct reached a climax in October 1961, with the detonation by the USSR of the 'Tsar Bomba', a device equivalent to 57 million tonnes of TNT – or, for comparison, 3,800 times more powerful than the weapon that razed the Japanese city of Hiroshima.[1] Until recently, the most perilous year in human history was 1985, when the world's nuclear arsenal swelled to a total of 61,662 atomic warheads[2] before declining under various treaties and modernisation programmes to a total of 12,700 devices in 2022 (see Figure 4.1).[3] What remains is, however, still far more than enough to extinguish the human species.

Despite the dwindling stockpile, the risk of nuclear holocaust has in fact increased. *The Bulletin of the Atomic Scientists*, which has maintained a vigil over the scale of the threat since it was founded by Albert Einstein and Robert Oppenheimer at the dawn of the nuclear age in the 1940s, set its Doomsday Clock at 100 seconds to midnight in 2021, and again in 2022. It said: 'Accelerating nuclear programs in multiple countries moved the world into less stable and manageable territory last year. Development of hypersonic glide vehicles,

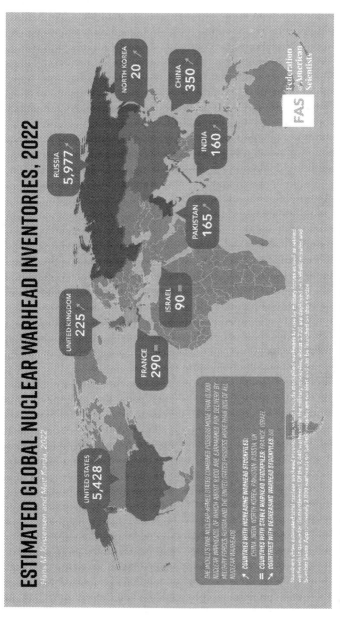

Figure 4.1 Global nuclear warhead inventories, 2022. Credit: Federation of American Scientists, 2022.

ballistic missile defenses, and weapons-delivery systems that can flexibly use conventional or nuclear warheads may raise the probability of miscalculation in times of tension.'[4] *The Bulletin* continues: 'An extremely dangerous global failure to address existential threats – what we called "the new abnormal" in 2019 – tightened its grip in the nuclear realm in the past year, increasing the likelihood of catastrophe.' It adds: 'Governments in the United States, Russia, and other countries appear to consider nuclear weapons more-and-more usable, increasing the risks of their actual use. There continues to be an extraordinary disregard for the potential of an accidental nuclear war.'[5]

The Bulletin argues that the threat is compounded by increased signs of instability and failure in Western democracies, epitomised by the assault on the US Congress in early 2021 by deluded Trump followers and a universal plague of false information that is confusing voters. At the same time, some nations are upping the ante in the nuclear arms race. China, for example, is threatening to double its existing nuclear arsenal: in late 2021 reports became public that the country is building up to 300 new nuclear missile silos in its western deserts.[6] These appear to be additional to its existing 350-warhead stockpile. The UK, India, Pakistan, and North Korea are also increasing and modernising their nuclear arsenals to make them stealthier, more accurate, faster, more concealable, more flexible – so increasing the danger of their use. Such expressions of 'nuclear nationalism', in an age of compounding mega-threats, are little more than a shortcut to species suicide.

Meanwhile, the emergence of a new generation of autonomous nuclear devices makes the likelihood of mass extermination by computer error or human misunderstanding even greater, in view of the many known near-catastrophes which have occurred since the nuclear age began.[7] Now, by replacing humans in the decision-chain, robots and artificial intelligence (AI) may make their own independent decisions to unleash Armageddon (see Chapter 10).[8]

The greatest single risk of human extinction among the 10 catastrophic threats that comprise our existential emergency is still nuclear war. However, the core issue is that conflict can originate with almost any one of them – with food shortages leading to international disputes over food, land, and water; in quarrels over dwindling fish, forest, energy, or mineral resources; in the unleashing of uncontrolled technologies such as cyber raids on national IT networks, banks, and even nuclear command-and-control centres; the release of novel man-made plague organisms; the almost universal brain damage and loss of IQ now being caused by the chemical flood; the tension and anxiety driven by worsening climate conditions; and in the manic tide of false information propagated by fools and malignant actors via the internet. The existential threat to humanity thus spirals out of the coming together of several of these mega-risks, culminating in nuclear conflict.

An instance of how mega-risks may compound into nuclear war is the long-standing animosity between India and Pakistan, chiefly over Kashmir, terrorism, and the waters of the Indus River which feed both countries at a time of growing climate stress. Even a relatively limited nuclear conflict between the two – 100–150 warheads of Hiroshima scale – is projected to kill 100 million people directly and 1–2 billion people worldwide as the resulting 'nuclear winter' would cause harvests to fail and food supplies to collapse all around the planet.[9] Such a disaster would almost certainly trigger further wars, some of them nuclear, as governments fail and atomic weaponry falls into the hands of political radicals, warlords, criminals, or religious extremists.

A second example is acute water scarcity leading to a food crisis in northern China, spilling the local population in all directions, including Siberian Russia: strategic think tanks fear such a development could precipitate a nuclear response. Another case is the Middle East, already the most water-starved and volatile region on Earth, where the acquisition of nuclear weapons by Iran could spark a regional arms race involving Israel and, potentially, Saudi Arabia.[10] In all these cases, the nine catastrophic threats pave the road that leads to nuclear holocaust – and all must now be regarded as primers in the explosive chain leading to civilisational collapse and human extinction.

The most hopeful moment of the nuclear age came in 2017 with the adoption by the United Nations of the Treaty on the Prohibition of Nuclear Weapons (TPNW).[11] Of those voting, 122 of the 193 nations who are members of the UN voted in favour of the treaty, with one against and one abstention. The rest remained tight-lipped or spoke against it. Worldwide, groups such as ICAN (the International Campaign to Abolish Nuclear Weapons), the Campaign for Nuclear Disarmament (CND),[12] Greenpeace,[13] and the Red Cross/Red Crescent[14] continue to urge all states to sign the treaty, which came into force on 22 January 2021. The nine nuclear states,[15] along with their 32 'nuclear umbrella' allies[16] and others they can coerce into siding with them, have remained stubbornly silent – although China, Russia, UK, the United States, and France reaffirmed in 2022 that a nuclear war 'cannot be won and must never be fought'.[17] None of them, however, has shown any serious intent or actually done anything purposeful towards total disarmament for decades.

As things stand, one in five of the world's nations continue to endorse a potential nuclear Armageddon and its possible outcome – human extinction. In addition to not having a plan for human survival, their default position embraces human destruction. Furthermore, on dozens of occasions, through active threat or technical or human error, the world has already come close to the brink of nuclear conflagration.[18] This is a terrifying history, of which most people remain ignorant. If humans do not wish to become extinct, the answer is plain: abolish the nuclear weapons machinery – now, in its totality, and forever.

The Solutions

1. *Outlaw and destroy all nuclear weapons and stocks of fissile materials (highly enriched uranium and plutonium).*
    Pathway: It is especially down to the citizens of the nine countries with the ability to destroy civilisation to put pressure on their governments to walk away from the risk of universal destruction, even though some of these countries are authoritarian. Without citizens taking prime responsibility, it is doubtful if national governments, politicians, and militaries ever will. It is up to the rest of humanity to support and encourage them in this cause in every way possible.

2. *Convert from uranium-based nuclear energy to safer systems (renewables, geothermal, etc.) less suited to producing weapons of mass destruction (WMD).*
    Pathway: The modern uranium reactor was originally designed to make the materials for nuclear weapons, even though it has since been adapted to produce electricity. Reactors still produce fissile waste, so their continued use is a threat to the human future and they should be eliminated for the same reasons we need to eliminate fossil fuels or pandemic diseases. Their use is chiefly promoted by the uranium mining industry and the nuclear power generators and their lobby groups.

The interdependence and co-promotion of the nuclear energy and nuclear weapons sectors is of increasing concern.[19] Nonetheless, the profitability of one small industry is no justification for imperilling all humanity. Only citizen demand and political pressure for clean energy can move the world away from such a risk. Radionuclides for use in medicine and science can be produced in much safer devices, such as cyclotrons.

3. *Outlaw and destroy all chemical and biological weapons and stocks.*

Pathway: There has been heartening international progress in this regard, proving it is possible to disarm.[20] But further action by citizens will be necessary to eliminate these WMD, coupled with greater international efforts to build trust, mutual dependency, and lower the risk of conflict between nuclear-armed states. This progress at least offers a working example of evidence-based WMD disarmament for nuclear states to follow. This action must occur not only in democracies but in all other forms of government, including centrally controlled, one-party republics and dictatorships. The citizens of these countries require all possible support from those in free countries for their efforts to persuade their rulers to give up these dreadful arms. Even in Western democracies, those of all political colours often support the retention of nuclear, chemical, and biological weapons, so that far greater citizen pressure to end political support for catastrophe is required.

4. *Develop stronger, more collaborative global surveillance of nations and groups that pose a potential risk of WMD terrorism.*

Pathway: Step up international co-operation in intelligence sharing on the nuclear threat, civilian oversight, and global surveillance through the UN of unstable nuclear states.

5. *Strengthen the global citizens' movement operating in all countries and societies to warn of the dangers of continued retention of WMD and a renewed arms race, and exert political pressure for their abolition.*[21]

Pathway: Use social media to reinvigorate and strengthen the worldwide disarmament movement.

6. *Make a stronger national and international investment in conflict resolution.*

Pathway: Strengthen current global institutions for peacemaking and resolving disagreements.

7. *Allocate a fixed percentage of the global military budget for addressing global challenges liable to lead to war.*

Pathway: Especially, earmark 20 per cent of global military spending to 'peace through food' – that is, ensuring a universal food supply adequate to reduce tensions that lead to conflict.

What You Can Do

- Understand that a nuclear inferno is a growing threat to you, your children, and to all of posterity. It exists 24/7. It is most likely to be the cause of human extinction. The fact that it has not happened in the last seventy years does not mean it will not happen. The risk is now greater than at any time since atomic weapons were invented.
- Actively support responsible citizen campaigns to ban nuclear, chemical, and biological weapons and demolish stockpiles in your own country (if it holds them) and globally. If your country is opposed to WMD, highlight this and encourage your leaders to step up their campaigning for a peaceful world. Support the Treaty on the Prohibition of Nuclear Weapons.

- Do not vote for politicians who do not commit to complete nuclear, chemical, and biowarfare disarmament and stock destruction. Lobby ceaselessly for a future for humanity free from the fear of nuclear war by:
  - engaging on social media with like-minded groups locally and worldwide to urge the elimination of all nuclear weapons;
  - asking your local, state, and national elected officials to endorse the campaign;
  - spreading the word via local media;
  - working with other groups and leaders in your community;
  - passing a 'no nuclear' resolution in your local government.

Sound advice on how to go about this is provided by Back from the Brink,[22] ICAN,[23] Ploughshares,[24] International Physicians for the Prevention of Nuclear War,[25] the International Red Cross and Red Crescent Movement,[26] and other anti-nuclear groups.[27]

- Share your views on the nuclear issue with friends and family. Spread the message.
- Do not invest in companies which either manufacture nuclear arms or their delivery platforms or which extract their precursor elements, such as uranium. Make sure your bank and pension fund does not invest in these companies either, as there will be no savings or retirement after a holocaust.
- Avoid belief systems which encourage discrimination or hatred for other groups or races, or extreme nationalism which may lead to conflict and nuclear war.
- Teach your children that, in the twenty-first century, with a precarious balance between human numbers and stressed resources and ecosystems, conflict is the road to ruin. There are no 'winners' – only losers.

Co-operation and mutual understanding, on the other hand, is the road to peace.
- Support the resolution of all conflict by peaceful means. Oppose the spread of all weapons, including those in civilian hands.

5 COOLING EARTH

The Problem

From his Melbourne living room, 72-year-old Dr Tom Beer, a retired scientist, watched aghast as Eastern Australia erupted in flame, with ash and mass death on the screen of his television. The inferno, the 2019–20 bushfire summer, sent shockwaves round the world over the whirlwind that humans are now starting to reap having ignored 50 years of warnings about global heating.

More than thirty years earlier, in what might have been the opening scene from a Hollywood disaster movie, Beer had been a lonely voice cautioning Australia, the world, and its governments of the consequences in a world-first report.[1] In it, he accurately predicted the combination of events that would ultimately generate holocausts from Siberia to the Canadian Arctic, California, the Mediterranean basin, Amazonia, Indonesia, and Australasia. Apart from a few fellow climate scientists, nobody paid much attention.

Scientists have been warning about human-caused climate change for two centuries. In 1824, French scientist Joseph Fourier calculated the Earth was warmer than its distance from the Sun indicated – and concluded its atmosphere was helping to store the extra heat. In 1896, Swedish chemist Svante Arrhenius observed carbon dioxide (CO_2) gas was especially good at trapping heat – and then realised that the massive CO_2 release from burning coal would cause the whole planet to warm. In 1938, Briton Guy

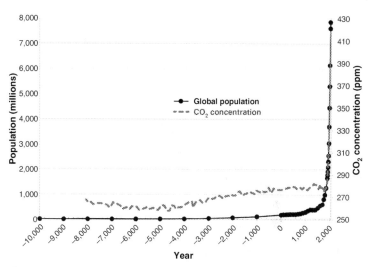

Figure 5.1 10,000 years of global population and CO_2 concentration. The growth in atmospheric CO_2 concentrations tracks the growth in human population. There is little doubt who is changing the atmosphere.
Source: Population data from Hyde 3.1 database; CO_2 concentration data from Dome C and Mauna Loa datasets. Credit: Peter Gleick, 2022, used with permission.

Callendar showed the planet had already warmed by half a degree over the previous half-century – and that this coincided with industrial growth and the growing use of coal. In 1958, US researcher Charles Keeling demonstrated a sharp rise in the CO_2 content of the atmosphere due to human activity. It had begun at 285 parts per million (ppm) in the 1850s; by the 1950s it was 315 ppm – and by 2022 it had risen to 422 ppm. Looking back at past atmospheric composition, researchers found that the recent outburst of CO_2 was quite exceptional – far beyond anything that had occurred naturally in the previous million years. Significantly, it closely parallels the explosion in human numbers (see Figure 5.1).

Equally exceptional were the rising temperatures and blistering heatwaves that had begun to afflict much of the

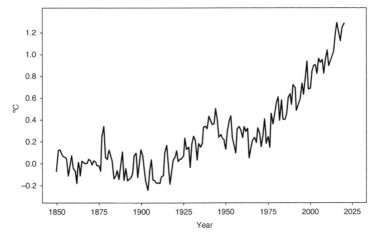

Figure 5.2 Global temperature increases 1850–2020, in degrees Celsius. Source: World Meteorological Organization, 2022.

planet. While the average increase worldwide was about 1.2 degrees Celsius (see Figure 5.2), temperatures in polar regions were up to 8 degrees hotter.

The first unambiguous warning to all humanity of the hidden dangers of fossil fuel burning and global warming was issued by the World Meteorological Organization in 1979. This was followed in 1980 by the *US Global 2000 Report* to President Carter which stated: 'Atmospheric concentrations of carbon dioxide and ozone-depleting chemicals are expected to increase at rates that could alter the world's climate and upper atmosphere significantly by 2050.'[2]

What has most surprised people is that, for the previous 11,700 years, the Earth's climate has been exceptionally stable with few large shifts in temperature. This stability allowed us to develop agriculture – and with that came cities, technology, organised religion, and social institutions. Without a stable climate to secure our food, *none* of this would have happened. Our civilisation is entirely climate-dependent.

That civilisation is now at risk, with the planet likely to cross the +2 degrees Celsius mark around the middle of the twenty-first century.[3] If current rates of man-made emissions persist, then by 2100 the Earth will be +3–4 degrees Celsius warmer than at the start of the modern age. Climate scientist Professor Will Steffen warns: 'At plus four degrees we are looking at the probable collapse of civilization.'[4] And Potsdam Institute Director Emeritus Hans Joachim Schellnhuber declared: 'I'm telling you that we're putting our children on a global school bus that will, with 98% probability, end in a deadly crash.'[5]

The reason for this is that our carbon emissions last a long, long time. Every time you start your car, part of the CO_2 it emits will still be warming the Earth up to a thousand years hence.[6] The CO_2 we release today may still be affecting the planet's climate 20 generations or more in the future: global heating is the ultimate form of child abuse.

Then there are the 'feedbacks' and 'tipping points' where the Earth system starts to heat itself – and humans can no longer do anything to stop it. Among the main tipping points are:

- loss of the Arctic sea ice in the northern hemisphere summer, increasing the heating of northern high latitudes through reduced reflectance of sunlight;
- collapse of the West Antarctic Ice Sheet, raising sea levels by 2–3 metres;
- drying out of the Amazon rainforest, releasing up to 60 billion tonnes of stored carbon (about 5 times what humans produce annually);
- slowdown of the Atlantic Meridional Overturning Circulation (AMOC) bringing drought, crop failure, and savage weather to western Europe and reducing rainfall over the Amazon basin;

- thawing of northern permafrost, releasing up to
  150 billion tonnes of carbon by 2100 (under high
  emissions scenarios) with ongoing emissions in the
  following centuries;
- collapse of the Greenland ice sheet, raising sea levels
  a further 7 metres;
- ecological collapse of the Northern Boreal Forests,
  releasing a further 100 billion tonnes of carbon;
- loss of 99 per cent of tropical coral reefs, leading to
  local collapse of ocean food chains.[7]

These feedbacks and tipping points operate on widely varying timescales. Some, such as melting of Arctic sea ice or conversion of the Amazon rainforest to a savannah, can occur within decades, while others, such as the loss of the Greenland ice sheet or large-scale thawing of permafrost, may take centuries to fully unfold. Tipping points and feedbacks can combine and reinforce one another to drive up global temperatures by a further 1–1.5 degrees on top of man-made heating of 4 degrees Celsius. These create the condition known as 'hothouse earth', in which large regions of the planet become uninhabitable.[8] Humans have an upper limit for survival in extreme humid heat of 35 degrees Celsius (95 degrees Fahrenheit) measured on a wet-bulb thermometer, where our ability to shed excess body heat by sweating fails and we start to 'cook' internally. Such extreme conditions have increased rapidly since 1980 and are predicted to become much more common by the mid-twenty-first century, affecting large regions such as the Middle East, South and South-East Asia, Central Africa, Australasia, and hot coastal areas worldwide.[9]

Under the tipping points scenario, the intricate machinery of the Earth's climatic balance, once disordered, cannot be repaired by any human action – but must run its course in a cycle which, we know from the geological past, may take hundreds of thousands, often millions, of years to return to 'normal'.

An excess of CO_2 in the atmosphere has accompanied every major and most minor extinction events since large animals first appeared on Earth around 700 million years ago (see Chapter 2, Figure 2.2). In the past, the main source of the CO_2 was outgassing from vast volcanic or tectonic events; now, it is human emissions from burning fossil fuels and clearing and farming land. The effect of past warming events caused by the excess CO_2 was first to melt the polar ice caps and then, by eliminating the ocean currents driven by the heat differential, to cause the oceans to stagnate and lose their oxygen, leading to a wipeout of sea life.[10] This stagnation then releases a further toxic gas, hydrogen sulphide, which poisons life on land, especially plants and trees, through acid rain.[11] It is this process which humans have unwittingly set in motion – and which has the potential to destroy us, along with most other species, in the longer term.[12]

However, the short-term consequences of a destabilised climate are already plain: increasing droughts and famines; floods, fires and more violent storms; water crises; many more heat deaths and diseases; millions of refugees fleeing ruined regions[13]; and an increase in violence and war.

'We are now very close to a fork in the road, where the Earth system goes out of control', says Will Steffen. 'This is an existential threat to civilization. No amount of economic analysis will help. Once we pass this threshold, there is no return. We will experience Hothouse Earth.'[14]

The Solutions

1. *Cease burning all fossil fuels by 2030 and replace them with renewables. No new coal mines, oil or gas wells, tar sands fields, or pipelines from now on.*

Pathway: The important thing to understand about energy use is that 66 per cent of the primary energy we produce is wasted,[15] mostly in the form of heat lost during production and use and also through energy-inefficient devices like motor cars, which use only about 20 per cent of the energy in the original petroleum to power them.[16] The opportunities to save energy are therefore colossal and, so far, widely neglected. Furthermore, humanity wastes around 40 per cent, or 1.3 billion tonnes, of the food we produce[17] – and food production is a massive consumer of energy for farming, transport, processing, storage, and delivery. Similar losses occur in the extraction of minerals, and our disposal of 11 billion tonnes of household and industrial waste. Our economic system is founded on the vast overproduction of unwanted goods and prodigal waste of resources. Saving ourselves from climate catastrophe involves making radical changes to our entire system – and every aspect of it, economics especially.

The Intergovernmental Panel on Climate Change's (IPCC) 2022 report[18] called for:

- major transitions in the energy sector, including substantial reduction in fossil fuel use, widespread electrification, improved energy efficiency, and use of alternative fuels (such as hydrogen);
- the reduction of emissions from industry by using materials more efficiently, reusing and recycling products, and minimising waste;
- large-scale emissions reductions and removal and storage of CO_2 across agriculture, forestry, and other land uses;
- worldwide electrification of the urban transport sector and domestic heating and cooling;
- accelerated measures to mitigate and adapt to climate change across all sectors and societies.

Detailed pathways and options for climate change mitigation are laid out in many UN, government, and scientific reports. They include strategies such as replacing worn-out coal-fired power stations with clean electricity generation; accelerated investment in renewables; carbon cap-and-trade systems; accelerated energy saving in industry, transport, and cities; novel energy technologies; reafforestation and revegetation of landscapes; recycling of materials; a reduction in intensive meat production, and so on.

2. *Reforest and revegetate half of the Earth's landmass as quickly as possible.*[19]

Pathway: This can be economically driven by carbon trading, but should follow the systematic approach outlined by E. O. Wilson in his book *Half Earth: Our Planet's Fight for Life*. The bulk of global food production should be shifted from rural landscapes to cities (where all the necessary water and nutrients are already available to produce food with a far lower carbon and resource footprint) and the deep oceans (see Chapter 8).

3. *Accelerate worldwide research and investment in clean, renewable energy.*

Pathway: Place renewable energy R&D on a 'war footing' worldwide to deliver new technologies in time to avert dangerous overheating of the planet. This is a pan-species collaborative effort that has to happen at global, bilateral, national, scientific, industry, and local levels. It encompasses all forms of green and renewable energy including solar, wind, wave, tidal, hydro, hot rock geothermal, green hydrogen, nuclear fusion, etc. It is imperative that every new energy technology – such as hydrogen[20] – should not cause any further catastrophic threats, and each must be carefully and independently assessed for risks before widespread adoption. Scientific precaution suggests that world energy supplies should be totally renewable by 2030.

4. *Develop global partnerships to accelerate the uptake of clean, renewable energy.*

Pathway: Use the model of successful government–private sector partnerships established for major infrastructure in recent decades to introduce clean energy at national, regional, urban, and local community levels. Subsidise the introduction of renewable energy at household and small community levels, especially in rural areas. Subsidise the electrification of transport globally.

5. *Progressively shift food production from fragile extensive agricultural systems to intensive local urban systems that use far less energy, transport, soil, water, and nutrients and cause less climate damage.* Increase research into meat alternatives such a biocultures. Shift global food culture from 'European' (meat and wheat) to 'Asian' (vegetables and sustainable seafood). Restore forests and savannahs to lock up carbon. Redesign farming and forestry systems to lock up carbon.

Pathway: Climate, water, and nutrient shortages will drive this transition, but the process can be accelerated by investment incentives and increased R&D. Provide economic incentives to farmers worldwide to lock up CO_2 by 'carbon farming', adopt regenerative farming methods, and manage 'rewilding'. Establish a global Stewards of the Earth programme to rewild half the Earth's land mass (see Chapters 8 and 15).

6. *Replace aviation and long-distance transport fuels, plastics, synthetic fibres, petrochemicals, and drugs made from petroleum with renewable carbon-neutral oil made from algae or green hydrogen and green electricity.* Replan and re-green cities and progressively replace private urban car ownership with efficient public transport powered by renewable electricity and 'active transport' (walking, cycling, scooters). Convert shipping to optimal energy systems including electric, wind-assisted, nuclear, and so on.

Pathway: Governments should provide incentives for industry to move urgently away from fossil fuels as a feedstock for transport fuels, chemicals, drugs, plastics, etc. in favour of natural substitutes (see Chapter 7). Educate consumers to drive market demand for non-fossil products.

7. *Drive a new phase of materials use efficiency in industry, so that nothing is wasted or allowed to pollute.*

Pathway: Introduce and promote concepts like 'cradle-to-cradle' manufacturing, zero waste, industrial ecology, green chemistry, and the circular economy universally around the world.

8. *Institute global and national economic reforms that swing the market in favour of clean, renewable energy; renewable food; and large-scale landscape renewal.*

Pathway: Since so many governments and politicians receive electoral funding and other incentives from fossil fuel companies, this will only happen through relentless citizen pressure, as well as legal and media exposure of corruption in government. Changes should include ending the subsidisation of fossil fuels, carbon taxes to redress market imbalances, re-localising production, stimulating investment in natural capital, restricting exploitative corporate investment in food, fishing, and forestry, and the outlawing of political bribes and donations by the fossil fuels sector. Adopt regeneration, recycling, redesign, retrofitting, and resilience in industrial policy.[21]

8. *Design and remodel cities and buildings to be carbon-neutral and climate-proof, to save energy, and to recycle both nutrients and water.* Create local heating and cooling networks. Exploit the vast opportunity for 'win–win' synergies between sustainable development and energy efficiency, renewable energy, urban regreening, reduced air pollution, and sustainable, healthy diets and activities.

Pathway: This is a challenge for urban planners, architects, engineers, and builders worldwide to conceive, plan, and build the clean, green, resilient metropolises of the future. Fortunately, urban governments are far more attuned to the needs of their populations that face various existential risks than are most national governments; the global competition to design the most renewable cities has already begun. It needs to go much faster, with ideas, technologies, and approaches shared by cities around the world at light speed via the internet.

9. *Reduce human population and consumption.*

Pathway: In the view of many eminent scientists the human population is now beyond, some say well beyond, the capacity of the Earth to sustain in the long run – especially at our present rates of material consumption and destruction. However, universal family planning availability can quite quickly and voluntarily bring back the human population to a sustainable level, once the need to do so is widely understood (see Chapter 9).

What You Can Do

- In the longer run, the greatest contribution any person can make towards fighting global heating is to have one child fewer than they may otherwise have chosen. This saves the world 58 tonnes of greenhouse gas emissions per person per year – whereas going without a car saves 2.4 tonnes and turning vegetarian saves 0.8 tonnes (see Figure 5.3 and Chapter 9).
- In the short run, the greatest contribution anyone can make to curbing climate change is to lower our personal consumption of *everything*, especially consumer goods; textiles and clothing; items of machinery and equipment; building materials; travel and transport; goods imported from far away; and

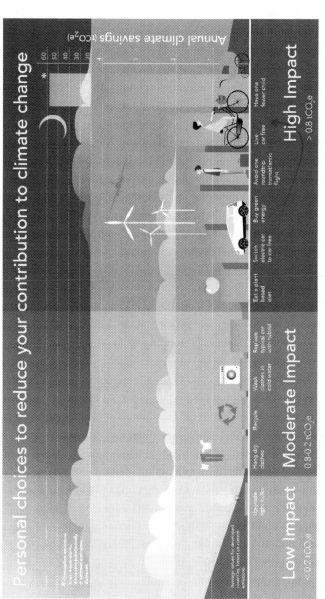

Figure 5.3 Greenhouse emissions saved per year from various individual actions. Data from Wynes S & Nicholas KA. The climate mitigation gap: Education and government recommendations miss the most effective individual actions. *Environmental Research Letters*, 2017, 2 (7). https://doi.org/10.1088/1748-9326/aa7541.

luxury foods – and recycle more and waste less. It is
our overconsumption of goods that is the chief driver
of climate change – and we have to get it under
control or else rob our grandchildren of their future
on a habitable planet.

- Use your power as an educated, caring consumer to
  send a price signal industry cannot miss by choosing
  only products produced with clean, renewable energy
  and which are 100 per cent recyclable.

- Do not vote for any politician who is not committed, by
  deed and word, to protecting your grandchildren by
  defending the climate.

- Take steps to reduce your own carbon footprint and
  help others to do so – much excellent advice is available
  from local government bodies,[22] universities,[23]
  charities,[24] climate researchers,[25] and concerned
  citizens.[26] The UN Environment Programme (UNEP)
  lists 10 tips to help fight the climate crisis.[27] Project
  Drawdown, a US-based non-profit organisation,
  specialises in providing sound, science-based advice on
  how to reduce your carbon footprint, including free
  online courses.[28] Typical advice for householders
  includes:

  - Cut food waste. Buy only food you know you will
    eat. Grow your own. Compost food waste and
    reuse.
  - Eat more veggies – replace some meat in your diet
    with healthy, protein-rich plants.
  - Walk more. Cycle instead of driving. Use public
    transport more, instead of private vehicles.
  - Switch to clean energy. Choose a new supplier
    based on how they generate and store power.
  - Insulate your home – improve your home's
    window, roof, and wall insulation.
  - Switch to energy-efficient light-emitting diode
    (LED) bulbs for home lighting.

- ○ Heat and cool smarter. Install heat pumps or smart thermostats.
- ○ Go solar. Invest in solar panels and batteries and solar hot water.
- ○ Drive electric or hybrid – make your next car an electric or hybrid vehicle.
- ○ Reduce and recycle – avoid single-use plastics and recycle as much as possible.[29]
- Teach yourself and your family about the lifecycle carbon content of manufactured goods and building materials – that is, how much carbon goes into mining, making, and transporting them as well as their actual use and disposal. Not everything is as green as it claims. Start by calculating your own carbon footprint, using a calculator such as the one provided by the Global Footprint Network,[30] or a commercial calculator such as Carbonfootprint.com.[31]
- Consider, for example: flying far less (at least until it becomes totally, truthfully carbon neutral), using public transport, walking or bicycling more, eating less meat, avoiding all forms of plastic, using clothing to keep yourself warmer/cooler instead of heating/air conditioning, planting more trees and food gardens, reducing home use of appliances.
- For those seeking even more good ideas to green their lifestyle, reduce their eco footprint, cut their carbon output, and be a better global citizen, a wide range of good books is available, such as:
  - ○ Philippe Boursellier, *365 Ways to Save the Earth*, Abrams, 2007
  - ○ Mike Berners-Lee, *There Is No Planet B*, Cambridge, 2021
  - ○ *The Sustainable Travel Handbook*, Lonely Planet, 2020
  - ○ Anita Vandyke, *A Zero Waste Life in 30 Days*, Penguin Random House, 2019

- ○ Karl Coplan, *Live Sustainably Now*, Columbia University Press, 2019
- ○ Paul Greenberg, *The Climate Diet*, Penguin, 2021.
- Support companies that demonstrate a strong climate ethos, both as a customer and as an investor. Avoid companies with investments or activities in fossil fuels, petrochemicals, plastics, large-scale land development, or palm oil and other destructive practices. Scrutinise your bank, pension fund, insurance firm and other financial services to see whether their investments are climate-friendly or climate-hostile – and make your decisions accordingly. Use an ethical investment guide to help you.[32] Do not let your retirement savings fund the destruction of your grandchildren.
- Favour fresh, locally grown food. It involves far less transport and refrigeration emissions, uses fewer chemicals and preservatives, is fresher and healthier, and creates more local jobs and economic activity in your community. If possible, grow more of your own food at home.
- Join the rapidly growing global movement to divest from the fossil fuel industry and oppose new mines and extraction. Besides dedicated divestment groups, religious bodies, banks, insurance and pension companies, universities and even governments are now removing funding from polluting industries and companies.[33]
- Join neighbourhood groups who work to lower local carbon emissions or sequester carbon in the soil. Share your own knowledge, experience, and advice with others. Hold 'kitchen table' conversations with your neighbours on the best ways to lower local carbon emissions – for example, by improving public transport services or generating electricity locally.
- Join online and global social media to share ideas for cutting carbon emissions, to express your views, and to

send a signal to governments and corporations all around the planet that its citizens want the dirty carbon era to end.

- Do not invest in, work for, or buy from any company that does not care (as shown in its emissions) what happens to your grandchildren.

6 CLEAN UP THE PLANET

The Problem

The Earth, and all life on it, is being saturated with human-emitted chemicals in an event unlike anything in our planet's 4-billion-year story. Every moment of our lives we are exposed to thousands of these substances. They enter our bodies with each breath we take, each meal or drink we consume, from the clothes and cosmetics we wear, our furnishings and cars, and the things we encounter every day in our homes, workplaces, and travel. They affect everyone on Earth, every single day.

Humanity emits around 220 billion tonnes (Gt) of chemical substances a year,[1] in a toxic avalanche that is injuring people and life everywhere on the planet. This is an act of pollution five times larger than our climate emissions but, despite its awesome scale, is almost never discussed by either government or industry.

The poisoning of our planet through man-made chemical emissions is, in all probability, the largest human impact on Earth – and the one that is least understood, measured, or regulated. It is one of the 10 catastrophic risks now confronting humanity.[2] This planetary mass pollution has mostly taken place in the last two generations. All previous generations inhabited a far cleaner world.

Recent assessments have identified more than 350,000 man-made chemicals in existence.[3] The US Department of Health estimates that up to 2,000 new chemicals are being released every year. The UN Environment Programme (UNEP) warns that most of these have never been screened for human health or safety.[4] In all, more than 2.7 billion tonnes of industrially produced chemicals are manufactured each year, and sales are forecast to double by 2030.[5]

The World Health Organization has estimated that 13.7 million people – 1 in 4 – die *every year* from diseases caused by 'air, water and soil pollution, chemical exposures, climate change and ultraviolet radiation', all of which result from human activity.[6] This is the largest mass homicide in human history – larger than either World War I or World War II, than massacres in the USSR and China in the mid-twentieth-century, than the Taiping Rebellion in the 1850s in China. On a typical day, more than 25,000 people die from poisoning or diseases caused by their chemical environment.

The toxic avalanche includes not only purposefully made chemicals but also the far larger release of waste substances emitted in the course of mining, farming, manufacturing, construction, transport, and land development:

- manufactured chemicals – 2.7 Gt per year;
- hazardous waste – 400 million tonnes (Mt) per year;
- plastics – 400 Mt per year;
- coal, oil, gas, etc. – 17 Gt per year;
- lost soil – 36–75 Gt per year;
- metals and materials – 18 Gt per year;
- mining and mineral wastes – 100–150 Gt per year;
- water (mostly contaminated with the above wastes) – 9 trillion tonnes per year.

Industrial toxins are routinely found in newborn babies, in mother's milk, in the supermarket food chain, and in domestic drinking water worldwide. They can be found in the blood of almost every citizen in the developed and developing worlds. They have been detected from the peak of Mount Everest (where the snow is so polluted it does not meet drinking water standards) to the depths of the oceans, from the hearts of our cities to the remotest islands and the poles.

These chemicals now move around the planet constantly in space and time, in cycles of absorption and rerelease called the *Anthropogenic Chemical Circulation*.[7] They travel on the wind and in water, in our food and drink, in traded goods, in wildlife and plants, in people and, though our genes, the damage they cause is being transferred to our grandchildren.[8]

The mercury found in the fish we eat is mainly fallout from the burning of coal, which contaminates the marine food web and increases every year. Our seafood and marine life are also contaminated with 14 Mt of plastics a year: these are made from petroleum and are now entering humans via the food supply.[9] There is global concern over the death of honeybees from 5 Mt of agricultural pesticides and the potential impact on the world food supply and on all insect life – as well as on the birds, frogs, and fish which in turn depend on insects.

An issue often overlooked by governments and corporations is that chemicals act in combination, occur in complex mixtures, and undergo constant change. They do not 'go away'. A single chemical may not occur in toxic amounts in one place – but combined with thousands of other chemicals it may contribute a much larger risk to the health and safety of the whole population and the environment that supports it.

Medical science is increasingly linking issues such as obesity, cancers, heart disease, and brain disorders such as autism, attention deficit hyperactivity disorder (ADHD), depression, underdevelopment, gender and reproductive difficulties, and declining intelligence to the growing flood of toxic substances to which humans are exposed daily. A significant development is the observed decline in human intelligence (IQ), since the mid-1970s.[10] While the causes are still unclear, the problem is not genetic and its occurrence coincides with the recent dramatic growth in emissions of neurotoxic petrochemicals.

Despite attempts to regulate chemical use, only 26 out of 350,000 man-made chemicals have been banned in the last 30 years.[11] Most chemicals have still never been tested for toxicity to humans or the environment. In many countries the petrochemical sector applies constant pressure on government to roll back existing laws that protect public and environmental health in the interests of profit.

The good news is that solutions to the mega-threat of global poisoning exist, but they require worldwide cooperation by consumers, government, civil society, and industry.

In my book *Earth Detox*, I proposed a new human right – a *right not to be poisoned*. It echoes the existing *right not to be tortured* and applies to every person in the world at this very moment. It is intended to raise awareness of the shocking death and injury toll from chemical emissions and to encourage industry, government, and citizens to take positive action to clean up our world. Without such a right, there will never again be a day in history when humans are free from man-made poisons – a day such as all our ancestors enjoyed.

We also need a global alliance of consumers and concerned citizens prepared to reject toxic products, or products made with toxic processes, and 'buy green' – so giving industry the economic signal to switch to green chemistry and other safer systems or products.

Communities the world over need to move as fast as possible to policies of 'zero waste', where nothing is discarded but everything is either reused or where toxic waste is made safe. The Earth has been poisoned because we, as consumers, send monetary signals every day to industry to make things as cheaply as possible. Almost all of this production involves the use of toxic chemicals or the release, at some point, of poisonous waste. Our unbridled demand for consumable goods and food takes no account of the damage to human health and life that uncontrolled chemical use entails, including the 14 million annual deaths and 602 million years of human life lost annually to disability.[12] So we are all, in a sense, getting away with murder.

If consumers demand safe, healthy, green products, and are willing to pay industry a little more to make them safely, we can cleanse our planet within a generation and save countless lives. My book *Earth Detox* outlines the path to achieving this.

The Solutions

1. *Form a global network of people and institutions to clean up the Earth.* Incorporate chemical clean-up into the Earth System Treaty.

Pathway: Hundreds of organisations dedicated to a cleaner environment exist already and can be found on the internet where they are now coming together in real time to share information and pressure governments for a cleaner world. They need our encouragement, support, and participation to go much faster and to join forces in a unified, global movement to cleanse the planet. An Earth System Treaty is a global compact establishing the basis for a healthy, habitable planet for humans and all life (see Chapter 14).

2. *Spread awareness, share knowledge, and help motivate industry to adopt clean production systems through consumers who 'buy clean'.* Help citizens to become informed clean consumers.

Pathway: This is already taking place via the internet and social media, but needs to accelerate and expand into schools and consumer education in all countries. Without consumers to drive this change, dirty, toxic chemicals will always be favoured over clean ones because they are usually cheaper.

3. *Implement a universal human right not to be poisoned.*

Pathway: Since everyone is being poisoned, it follows that there is a universal need for such a right. Consumer groups, lawyers, and human rights bodies can make the case for the inclusion of this right in the Universal Declaration of Human Rights (UDHR) (see also Chapter 14).

4. *Replace all coal, oil, and other fossil fuels with clean energy and with clean feedstocks for industry.*

Pathway: The elimination of fossil fuels for the protection of the Earth's climate can, at one and the same time, also eliminate the main source of toxic pollution of humanity and life on the planet. Since most toxic industrial chemicals are made from petroleum, coal, or gas, eliminating them for climate reasons will deliver a second major 'win' for humanity by removing a source of poisoning that kills and disables millions every year (see Chapter 4).

5. *Eliminate the use of all known toxic substances from the food chain, water supplies, personal care products, home goods, and the wider environment.*

Pathway: Stronger citizen pressure is required to compel governments to act on known toxins. Most of them know what is bad for us but, under industry pressure, do little or nothing to prevent it. Citizens need to become informed consumers, shun all products containing or releasing known toxins, and expose weak regulation that endangers lives.

6. *Establish a worldwide inventory of all manufactured chemicals, prohibiting the use of all those not listed.* Independently verify the safety and toxicity of all chemicals in use in industry and wider society. Ban those which are found to be unsafe. Require all new chemicals to undergo human and environmental safety testing prior to release. Press for the prevention of disease, as opposed to using chemical 'cures' for diseases that are often caused by chemicals.

Pathway: It is essential that a global intergovernmental body for chemicals, akin to the Intergovernmental Panel on Climate Change (IPCC), is established. This will register all chemicals produced worldwide, assess the impact of chemical emissions on human health and planetary systems, and oversee toxicity testing.

Citizen pressure is needed to refocus the medical system on the needs of citizens for disease prevention, not just 'cure'. Much of the modern medical profession acts as a sales force for the global pharmaceutical industry, which prefers the treatment of diseases to be long, costly, chemicalised, and difficult, so as to sell more drugs, more profitably.

7. *Train all young chemists, scientists, and engineers in the social and ethical responsibility to 'first, do no harm'.*

Pathway: all major scientific disciplines should introduce an oath or pledge for new graduates not to use their science in any act or technology that may cause harm to humanity or the Earth that sustains us. This is already part of the Hippocratic Oath[13] taken by medical doctors on graduation from university. There is no ethical reason for universities to train other science graduates to cause harm; thus a universal Hippocratic injunction is desirable to inculcate harm prevention in all professional codes from the start of an individual's career.

8. *Empower industry to make profits ethically, by producing clean products that do no harm.*

Pathway: The quickest way to clean up industry is for consumers to send economic signals that they value clean products – and the quickest route to that is to educate consumers about what is safe and what is toxic. Many organisations are already doing this, but the learning needs to go farther, faster, and wider, via the internet.

9. *Reward industries that adopt approaches such as green chemistry, product stewardship, and zero waste with our patronage and support.*
Pathway: Improve consumer education about what is safe and what is not.

10. *Implement mandatory toxicity testing of all new industrial substances, major waste streams, and mixtures and daughter products of chemical waste.*
Pathway: This requires both national and global regulation, led by bodies such as the Stockholm Convention and reinforced by government environmental protection agencies, as well as parent and consumer bodies.

What You Can Do

• Learn to distinguish between foods and consumer products that contain toxins or are made using toxic processes: try to exclude them from your home, your work, and your life.[14] Minimise your intake of pesticides:
  ○ Buy organic food and clothing.
  ○ East fresh food, to avoid preservatives.
  ○ Thoroughly wash all fruit and vegetables (even organic).
  ○ Grow your own vegetables. Bake your own bread.
  ○ Peel vegetables or remove the outer layer of leaves.
  ○ Trim visible fat from meats (fat contains far more pesticide residue than meat).

- ○ Cook meat and chicken thoroughly to break down toxic chemicals. Avoid charring.
  - ○ Consume a wide variety of foods to limit and spread your intake of antibiotic-resistant bacteria, as well as hormones, pesticides, and preservatives used in industrial food production.[15]
- Do not feed children on foods containing substances made from fossil fuels (e.g. colouring or pesticides). Eliminate volatiles,[16] plastics, heavy metals, and nerve poisons from the family home, especially in furnishings, electronics, personal care and cleaning products, clothing, bedding, plastic toys, cutlery, etc. Reduce your intake of food containing chemical preservatives by eating fresh food (i.e. eat less ham, bacon, salami, crisps, manufactured bread, and other highly processed foods).
- Use your power as a consumer to penalise companies that emit toxins by avoiding their products – and reward those which produce clean, safe, recycled products and use clean processes by buying their goods.
- Participate actively in parents', citizens', and consumer groups dedicated to learning how to live more safely and healthily so as to clean up the Earth – or just your local community.
- Educate your children to choose wisely among products and services, based on their personal and universal health impact. Empower children to educate us.
- Use your power as a voter to compel governments to take their duty of care towards children and future generations more seriously, and to strengthen regulation and oversight of all chemical emissions, deliberate or not.

7 PREVENTING PANDEMICS

The Problem

In December 2019 a cluster of 41 cases of a mystery infection was detected in Wuhan, China.[1] In less than two years the novel plague blazed around the Earth, infecting over 500 million people and killing 15 million of them.[2]

Nobody should have been surprised. Since 2000, there have been seven new pandemics – epidemic diseases that spread internationally – SARS, swine flu (H1N1), MERS, Ebola, Zika virus, Covid-19, and monkeypox. Meanwhile, the HIV/AIDS pandemic, which began in the 1970s, continues to rage, claiming over 40 million fatalities worldwide.[3] In all, these biological ambushes may have taken the lives of as many as 50 million humans, or more, since the start of the present century, putting them on a par with the two world wars for sheer deadliness.

Pandemics are nothing new, as Table 7.1 indicates. Indeed, they are a prominent element in human history, since we first began to urbanise, travel, trade, and invade in large numbers. The three plagues that undid the Roman Empire probably arrived in the Mediterranean region via the Silk Road from Asia. The Black Death, another Asian import, was a consequence of the Mongol invasion of Europe (and gave rise to the concept of quarantine). The 'Spanish flu' of 1918 first appeared in the United States and was then probably spread globally by soldiers returning from World War I.

Table 7.1 *Recent major pandemics in human history: The frequency is growing*

Date	Plague name	Cause	Source
165–266	Antonine/ Cyprian	Measles	Travellers from Asia
541–543	Justinian	Yersinia pestis	Fleas from wild rodents
1347–1351	Black Death	Yersinia pestis	Fleas from wild rodents
1817–1824	1st cholera pandemic	Vibrio cholerae	Contaminated water
1827–1835	2nd cholera pandemic	Vibrio cholerae	Contaminated water
1839–1856	3rd cholera pandemic	Vibrio cholerae	Contaminated water
1863–1875	4th cholera pandemic	Vibrio cholerae	Contaminated water
1881–1886	5th cholera pandemic	Vibrio cholerae	Contaminated water
1885–present	3rd plague	Yersinia pestis	Fleas from wild rodents
1889–1893	Russian flu	Influenza A/ H3N8?	Birds
1899–1923	6th cholera pandemic	Vibrio cholerae	Contaminated water
1918–1919	Spanish flu	Influenza A/H1N1	Birds
1957–1959	Asian flu	Influenza A/H2N2	Birds
1961–present	7th cholera pandemic	Vibrio cholerae	Contaminated water
1968–1970	Hong Kong flu	Influenza A/H3N2	Birds
1970s–present	AIDS	HIV	Apes, monkeys
2002–2003	SARS	SARS-CoV	Bats, palm civets
2009–2010	Swine flu	Influenza A/H1N1	Pigs
2015–ongoing	MERS	MERS-CoV	Bats, camels
2019–ongoing	Covid-19	SARS-CoV-2	Bats, pangolins

In 2009, scientists concerned at the apparent upsurge in the rates at which animal and bird diseases were entering the human population produced a list of 1,400 organisms in animals with the potential to cause disease in humans, should they cross. They concluded that 500 of these were capable of achieving human-to-human transmission and 150 of causing an actual pandemic. At least 87 were recent transfers from wild animals to humans.[4]

However, recent science suggests even this may understate the problem. On average, two new potential plague organisms are crossing into humans from animals every year, though not all cause epidemics. These 'spillover' events are driven by activities such as land clearing and forestry, wildlife trade, hunting, and risky agricultural practices (e.g. avian flu outbreaks have been linked to Chinese farms containing both ducks and pigs). Climate change is also pushing birds and animals to move to new places.[5] All of these bring humans into closer and more frequent contact with wild animal hosts of disease, creating fresh opportunities for disease-causing organisms to transfer between species. At the same time, explosive growth in world demand for meat has led to a dramatic expansion of livestock industries,[6] creating an ideal reservoir for the incubation of diseases, new and old, such as tuberculosis, influenza, rabies, salmonella, and toxoplasmosis. In 2022, the world held 1 billion cattle, 680 million pigs, and 33 billion chickens – to feed 8 billion humans.

When superficial factors are stripped away, however, the principal driver of today's pandemics is human overpopulation – an unpalatable truth which government and UN officials are reluctant to voice aloud. As our numbers swell, the opportunity for the exchange of infectious diseases, both new and old, grows in proportion. Furthermore, overpopulation leads to increased population density – more humans per square kilometre to spread new diseases around; today's megacities are a major nursery for pandemics, both through overcrowding and through their many facilities which are perfect for transmitting disease – such as schools, childminding centres, night clubs, sports arenas, supermarkets, wet markets, shopping centres, poorly ventilated offices, transport hubs, and other services that pack people into close contact with each other. Amplifying all this is the massive upsurge in global travel – air, sea, and land – prior to the Covid-19 outbreak, providing a designer mechanism for the efficient worldwide distribution of diseases new and old.[7] Part of the overpopulation trend, researchers say, is a general increase in sexual promiscuity accompanying the emergence of a globalised society and globalised mores, as sex provides a primary avenue for the spread of infectious diseases.[8]

The intense focus that governments, the media, and society more widely has placed on the immediate culprit – the Covid-19 virus – leads us to largely overlook the fact that most pandemics are, in reality, an artefact of human behaviour. They are not the fault of the organism, which never set out to infect us in the first place – they are caused by humans, who first expose themselves to the natural reservoir of the new plague and then share it with one another by all possible means.[9]

Little wonder, then, that in 2020 Dr Mike Ryan, head of the World Health Organization (WHO) Emergencies Programme, cautioned that Covid was 'not necessarily the big one' and 'the next pandemic may be more severe'.[10]

As German Minister for Foreign Affairs Heiko Maas, said: 'COVID-19 has painfully reminded us that the health of humans, animals and the environment around the world is closely connected: Nobody is safe until everybody is safe. This is what we have to bear in mind to prevent future pandemics.' In December 2021, WHO Director-General Dr Tedros Adhanom Ghebreyesus announced a plan for a world accord to strengthen pandemic prevention, preparedness, and response.[11]

The Solutions

1. *Provide the services that will allow women everywhere to voluntarily reduce the human population.* As diseases that travel around the world, pandemics depend on large human populations to sustain them and to develop new subtypes. There is good evidence that a smaller population will also lower the risks of new disease emergence and transmission. There are many sound reasons to reduce human numbers, but the issue is so fraught with emotion and vested interest that most governments and the UN seldom dare to voice it for fear of a backlash. However, as human numbers begin to abate, the danger of pandemics will also start to decrease. Lowering fertility voluntarily is a personal decision that contributes to a healthier, safer world (see Chapter 9).

2. *Travel less.* Pandemic diseases depend on people travelling, and the transport of goods, to get around the planet. It follows that restrictions on travel and stronger biosafety and quarantine rules are needed. These are likely to be unpopular among travellers who place their own pleasure and convenience above the health and lives of others, but self-gratification does not justify homicide. The Covid-19 event has shown that there are other ways to 'travel' – virtually and via the internet or by exploring our home region first.

3. *Redesign cities.* By reason of their population density and dubious hygiene, cities have always been ideal breeding grounds for epidemics and pandemics. They need to be redesigned for a pandemic world to lower opportunities for transmission and to promote health, rather than

undermine it. This will involve more intensive use of social distancing, masks and protective clothing; virtual gatherings; and stronger social rules – such as those that discouraged coughing and spitting in public in the early twentieth century to prevent the spread of tuberculosis.[12] The progressive elimination of air pollution, as use of fossil fuels declines, will remove one of the main conditions that undermines the health of citizens and makes them more vulnerable to infectious diseases. The redesign of buildings and their air supply systems is imperative. Public facilities, such as schools, theatres, and sports arenas all need to be redesigned for distancing, fresh air, and reduced physical contact. By producing more of their own food using recycled nutrients and water, cities can dramatically reduce the pressure to clear more wilderness and forests.

4. *Stop land clearing.* If we wish to prevent novel diseases from entering our species, there must be worldwide agreement to end the destruction of rainforests and wilderness, which compels wild viruses to take refuge in humans. Furthermore, the global restoration of forests, grasslands, and wilderness (see Chapter 8) will also help to dilute the incidence of contact between humans and wild viruses that leads to spillovers. We need to strengthen international and local co-operative efforts to restore soils and water, replant forests, and repair landscapes and oceans, as proposed in Stewards of the Earth (see Chapters 2, 8, and 14) and with programmes such as the UN-REDD scheme.

5. *Inform and educate.* Global awareness and education is needed to teach people that new diseases usually come out of ruined ecosystems, and those environments are being ruined by our own economic signals as consumers. Consumer economics thus drives the growing risk of pandemics – and, equally, offers a solution through informed consumers, ethical corporations, and sustainable industries. We should build price signals into food and

other resource-hungry products that stimulate reinvestment in natural capital and avoid its destruction.

6. *Early warning.* Improve global and national early warning systems for new pandemics. WHO has been calling for this for years, but Covid-19 still caught the world unprepared. Apparently, national governments rarely learn the lessons of history, or else forget them very quickly.

7. *Ban deadly scientific experiments.* In the light of evidence that Covid-19 may have started with scientific engineering of a wild virus carried out in the course of vaccine research, all such research ('gain of function' or GOF) should be universally banned.[13] Scientific efforts to find new viruses in the wild and import them to centres of population should also be strictly and independently overseen.

8. *Strengthen health care.* Recognise that plagues often start or spread in disadvantaged regions of the world, and an imperative aspect of plague prevention is to build up the medical and health care capability of all countries, following the saying 'Nobody is safe until we are all safe.' There must be increased global efforts to strengthen core public health infrastructure, including water and sanitation systems and increased information sharing and disease monitoring, and well-laid plans should be put in place to move rapidly to extinguish sparks that could lead to a pandemic.

In a pandemic, local health care facilities and the health system overall can easily be overwhelmed, so it is essential to plan well ahead, to stock vaccines, drugs, and essential equipment, and to build up auxiliary health care services ready to respond to unexpected high demand. The principles of containing a pandemic are well understood and ought now to become automatic in every society.

9. *New drugs.* Fund a major global effort to develop new antibiotics and antivirals, preferably in the public sector.

Pathway: WHO and world medical authorities are already working on this, but the emergence of new antibiotic-resistant strains of old diseases makes it increasingly urgent. This needs to be coupled with predictive systems for ecosystems facing profound stress, where new pathogens are most likely to erupt.

10. *Destroy all stocks of extinct plagues, such as smallpox.* Outlaw the scientific development of novel pathogens with potential to harm humans. Ban dangerous scientific experiments such as GOF, which can create novel or more deadly plague organisms.[14]

Pathway: Like nuclear weapons, this pathway is blocked by the refusal of militarised nations, chiefly the United States and Russia, to disarm. Only citizen and voter action can compel them. GOF research is off the leash worldwide and must be brought under ethical control by society.

11. *Business, whether local or global, needs to better understand its role and responsibilities in pandemic prevention.* This applies especially to sectors such as air travel, tourism, agribusiness, land development, construction, and urban planning. All businesses depend on having healthy staff and customers. It is not in their financial interests to shirk their responsibilities for good pandemic planning, which should be a required benchmark in all annual reporting.

What You Can Do

- Consider having one child fewer (see Chapter 9) so that all children may lead a healthier, more disease-free life, leading to a healthier world.
- Avoid overcrowded venues and transport hubs as much as possible, realising that you are constantly sharing other people's breath and diseases and helping to spread them.

- Meet online or by phone instead of face-to-face wherever practical.
- Travel far less; read travel warnings; obey all quarantine and biosafety rules. Wear a mask when travelling, especially internationally.
- Make sure you are vaccinated against all the infectious diseases you may encounter. This is not only a matter of personal protection – you are helping to protect your family and fellow citizens. Risks from vaccines are vastly lower compared to risks from infectious disease and, in a pandemic, the unvaccinated generally die first and in the largest numbers.[15] There are 25 vaccine-preventable diseases, with another 15 new vaccines under development.[16]
- Wear masks and other protective clothing when asked by health authorities to do so, out of respect for the rights of others and the need to curb the spread of epidemic disease in the community – not just for personal protection.
- Do not share unsubstantiated stories about infectious diseases or vaccines either by word of mouth or on the internet.[17] Refer only to trustworthy sources such as WHO, your country's health authority, medical universities, and hospital staff. As is the case in war, careless talk costs lives – and the spreading of unsupported rumours about disease and its supposed cures can have fatal outcomes for those who are deceived by them. Also, the spreading of unfounded claims about vaccines and other public health measures can kill other people, and is an act of intentional homicide.
- Avoid buying any products which contribute to the global destruction of rainforests, such as those containing palm oil, tropical timbers, soybeans, chocolate, rubber, and beef; use sustainable and organic produce labels and logos as a guide to what is

safe, healthy, and sustainable to eat. Avoid buying
wildlife products and illegally smuggled pet animals.
Use your spending power to oppose deforestation and
encourage reafforestation. Research what goes into the
food, clothing, and consumer products you buy – and
make a clear decision to favour those that are sustainable, as distinct from those likely to lead to future
pandemics.

While many people may complain about being asked to
travel and mingle less or to vaccinate and wear protective
clothing, the public worldwide needs to understand that
*all pandemics derive from human behaviour – and changing that
behaviour is the only thing that can effectively prevent them.* We
cannot rely wholly on medical miracles. Also, what may
have been relatively safe behaviour in the 1950s, when
there were 2.5 billion people on the planet, has become
far riskier in the 2020s, in a world of 8 billion people
(including 1.4 billion tourists). It will be deadlier still if
there are 10–12 billion humans. To avoid ever larger,
more deadly pandemics, we need to adapt our behaviour
until we have managed to reduce our numbers to a safer
level – or disease will do it for us.

8 RENEWABLE FOOD

The Problem

By far the most destructive implement on the planet is the human jawbone. Each year, in the course of devouring 8.5 trillion meals, our jaw helps to dislodge up to 75 billion tonnes of topsoil, swallows 7 billion tonnes of fresh water, generates nearly a third of humanity's total greenhouse gas emissions, and distributes 5 million tonnes of specialised poisons.[1]

It fells forests and clears jungles, empties oceans of fish, destroys rivers and lakes, sterilises landscapes, spreads deserts, and helps to blanket the planet in a toxic shroud. It has played a central role in eliminating three-quarters of the world's large animals and birds and is the principal driver of the mass extinction of other species.

To put it bluntly, humans are in the process of consuming their planet, which – if you consider the matter even briefly – is not a good lookout for the fate of our own species.

Farming has been our principal source of food for over 5,000 years – and today's cities would not exist without the ability and skills of farmers to feed them. However, farming only came about because of a very special period in the Earth's climate history, the Holocene, when the climate was far more stable than usual (Figure 8.1). That unique period of stability has now ended, thanks to man-made climate change. The future climate will not only be

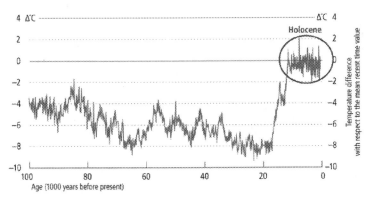

Figure 8.1 The Holocene (far right) was a period of exceptional climatic stability that enabled agriculture and cities to arise. Credit: Steffen W, based on data from Vostock Ice Core, 2020.

hotter, it will be more variable, violent, and far less suited to agriculture – with savage droughts, fires, and heatwaves followed by raging floods, crop-wrecking storms, and pest and disease outbreaks. A warmer planet means greater evaporation from the oceans, which in turn leads to heavier local dumps of rain, hail, and snow, all of which can ruin farm production. We have already had a foretaste of this – but scientific projections agree it is likely to get many times worse.[2] This will send shockwaves along the supply chains that now deliver much of the world's food to large cities, supermarkets, and retailers, triggering periodic food shortages and soaring prices.

The havoc wreaked by climate change will combine with continued loss of topsoil, dire shortages of water as mega-cities steal it from farmers, growing scarcity of fertilisers, and widespread chemical poisoning to ensure that farming, as a technology, will not be able to feed 10 billion people struggling to survive on an overheating planet in the 2050s and beyond. Farming, a perfect food solution for the stable Holocene, is now disastrously vulnerable in the tumultuous Anthropocene. *In short, humanity needs a new way to produce food – and fast.*

Take heart. Solutions are at hand. They are practical, involve little or no new technology and, what is most important, are mostly completely affordable – the money to implement them already exists. So do the people and the skills. However, as you may imagine, it involves a renewable food revolution greater even than the renewable energy revolution now sweeping the planet. But it is equally promising and feasible. It offers fresh opportunities, more jobs, healthier food, and new hope.

The first thing that everyone who eats needs to understand is that the present food system, perfectly adequate for the twentieth century and, indeed, the primary cause of the human population explosion, is not sustainable in the twenty-first. Apart from growing vulnerability to climatic impacts, modern industrial farming systems are destroying the very soils, waters, and ecosystem services they depend on at such a rate that major food system failures will be unavoidable in coming decades, starting with water crises in the 2020s and beyond. Just because our Bronze Age food system has served us well for 6,000 years, this does not mean it will work for 8–10 billion people in the hot, stressed, resource-depleted, ecologically impaired world of the late twenty-first century.

Food failures, we know from history, nearly always lead to wars and mass refugee upheavals. Only this time, thanks to globalisation, they are liable to be global in impact. And war is almost as bad for ecology as food production. This process is already under way, with one-third of a billion people leaving home each year in search of new lives in countries which appear to them more stable and food secure, giving rise to conflict over dwindling living space.

Thus, in developing a new food system, we must also find a way to curb the human appetite for war.

In my book *Food or War*, I traced the connection between food and conflict through human history, explored the food driver in recent and existing conflicts, and identified nine regions of the world which are at high risk of conflict in the foreseeable future – conflicts which range on a scale from riots and government failures to thermonuclear war. My aim was to show that the link between food and war is inexorable – but that it can be broken. And that having plenty of food for everyone is the most underrated, under-recognised, and badly needed 'weapon of peace' in the world today.

So, how do we achieve sufficient food for all of humanity, to take us past the peak in human population and back down to the sustainable level of 2.5 billion that existed when I was born (and towards which the world's women are now steadily leading us) without laying waste to the entire planet either agriculturally or militarily?

There are basically three pillars to a sustainable global food supply, each able to supply roughly a third of humanity's food needs:

1. Regenerative farming and grazing, globally, can restore ecosystem function over an area of about half of the planet which is presently being farmed or grazed. This will not only protect the rural landscape but also use minimal inputs of chemicals and fertiliser, lock up far more carbon, and treat animals more kindly.

2. Urban food production, in which all urban water and nutrients are recycled in a 'circular economy' into climate-proof food, using mostly indoor techniques from hydroponic, agritectural, and aquaponic to 'cellular agriculture' systems like synthetic meat, milk, fish, and eggs.[3]

3. Redoubled marine aquaculture, especially in deep-water ocean culture and algae (seaweed) farming or

water-cropping. This new source of vegetable and protein food will replace wild-harvest fisheries and some large-scale cropping on land. Because sea farms are three dimensional, they can produce a lot more food per cubic kilometre of ocean than from a square kilometre of land – 10 to 100 times or more. They thus have far less impact on the planet.

There is a lot more to each of these than I can explain in this short chapter, so please bear with the argument. Suffice to say that there are thousands of scientists, farmers, companies, and innovative technologists all round the world already pioneering these techniques, hammering out the flaws, and investing billions of dollars in 'renewable food' ventures aimed at a safe, healthy, sustainable diet for all. These techniques are not only able to supply a healthier, safer diet to our overfed societies – but also renewable food supplies to countries and regions suffering from hunger and war.

Furthermore, there is a dramatic opportunity for everyone to eat better. So narrow is our present industrial food base that we presently eat fewer than 300 (i.e. less than 1 per cent) of the 30,500 edible plants so far identified on Earth.[4] This means we have barely begun to explore our planet in terms of what is healthy, safe, and sustainable to eat. There are so many new foods, known only to indigenous societies, for people at large to discover.

In his book *Half Earth*, biologist E. O. Wilson argued that we need to set aside about half the planet for other life if we are to avoid mass extinction and an ecological collapse that will imperil our own future. In *Food or War*, I show how this may be achieved – by re-wilding half of the world's presently farmed and grazed lands, in all continents, under the stewardship of former farmers (whom the industrial food system is evicting anyway) and indigenous peoples – a scheme titled '*Stewards of the Earth*'. On Wilson's calculus, this should spare around 86 per cent of the species presently destined for man-made midnight.

Is this affordable? Yes. The funds to make it happen already exist. By diverting just 20 per cent of the global arms budget of US$2.1 trillion (i.e. US$420 bn/yr) to new ways to produce food, we can feed everybody – while reducing the risk of war by about two-thirds.

If you study the map of wars in the last generation or so (Figure 8.2), most have broken out in places where food, land, and water are insecure – and people are fighting over them, usually on cultural, religious, or political grounds. However, the same map also reveals that there have been very few or no conflicts in regions where food, land, and water are secure and there are few disputes over them. Furthermore, in areas which have suffered long periods of conflict, re-establishing secure food for all has often proved to be a key ingredient in the peace process.[5]

To those who may argue that defence budgets can never be cut, this is simply not true. Between 1990 and 2005, world military spending fell by up to 40 per cent, as Cold War fears receded. This demonstrates that it is perfectly feasible to reduce defence budgets without risking war. Indeed, at the time they were cut, there were fewer conflicts than ever around the planet. The simple way to a safe, sustainable, peaceful human future is to encourage all governments to invest more in renewable food systems – and less in weapon systems.

Such is the insatiable power of the human jawbone that renewable food not only holds the key to health, peace, and plenty for all, but also to ending the Sixth Extinction and regenerating a fairer, greener, safer Earth.

The solution to global food security for all rests in totally redesigning the world diet and food system around regenerative farming, city production, and deep-ocean aquaculture. Such a system can feed humanity through the peak in

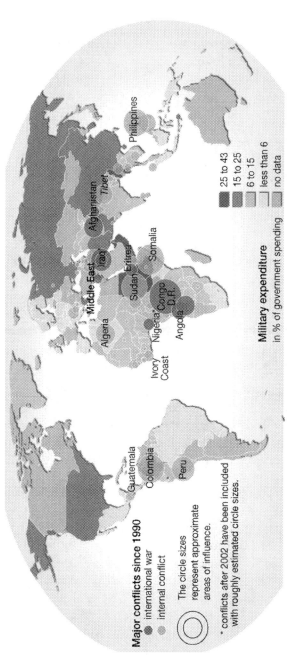

Figure 8.2 Map showing recent major conflicts since 1990 and military expenditure as a percentage of government spending. Credit: OCHA/Reliefweb.

its population in the second half of the twenty-first century and into a more sustainable future.

The Solutions

1. *Develop sustainable urban food production using recycled water and organic waste in all of the world's cities, especially in food deficit countries, to supply up to one-third of total global food needs.*

Pathway: Climate instability will drive much of today's farming indoors, while water, soil, and fertiliser shortages will also encourage the trend to renewable food systems. However, urban planning (to recycle both water and nutrient waste back into food production), investment incentives (by governments anxious to avoid climate-caused food crises), and more R&D into urban food systems, including biocultures, are essential. This entails building local fresh food production systems and networks, so reducing the climate and environmental penalties of long-distance food transportation.

2. *Convert the current world agricultural system to regenerative farming, at lower intensity, greater climate resilience, and fewer chemicals.* Rewild up to half the land mass to protect the environment, its water, soils, biodiversity, carbon lock-up, and other services. Regenerative agriculture can easily supply a third or more of the world's food needs.

Pathway: Many farmers worldwide have already embarked on this path, seeking to put their farms on a far more sustainable footing than the current industrial system permits. But they need much more help from society and governments. Regenerative farms restore soil and water quality, lock up carbon, protect the surrounding environment, and operate more effective and sustainable combinations of crops, pastures, and livestock. They dramatically reduce human pressures on the natural landscape and its resources. Scientific research and extension to support regenerative farming methods is urgently needed: at present much science is devoted to unsustainable practices. More farmer 'heroes of sustainability' are needed to champion the practicality and benefits of these methods.

3. *Develop deep-ocean farming of marine fish, animals, and plants on a global scale, to replace wild harvest of fish and supply up to one-third of the world's food needs.*

Pathway: Deep-ocean aquaculture avoids the problems of pollution, overcrowding, disease, and pesticide overuse that have beset coastal fish farms. 'Waste' nutrients from the food system can be used to grow marine algae (seaweed and microalgae), which in turn can be used to feed fish and other livestock for human consumption, as well as contributing to the human diet. Because they are three dimensional, deep-ocean farms can produce far more food per hectare than land-based farms, without overstocking. Excess nutrients are carried away and dispersed by ocean currents.

4. *Raise the coming generation of humans to value and respect food far more than we do today.*

Pathway: This can be achieved by introducing a Year of Food (see point 9) in every subject in every junior school on the planet, and encouraging the food industry to help educate consumers by publicising sustainably produced foods. The knowledge and means to do this already exist. It is a matter of persuading education providers that good health, sustainability, peace, and human survival are profound and essential aims of education. A school farm or renewable food garden can be set up at very little cost, with support from local businesses and community donors, to provide 'hands-on' learning and experience of the benefits of sustainable food production, preparation, and consumption.

5. *Replace the current waste of around 40 per cent of world food with a system that recycles all nutrients back into food production, especially in cities.*

Pathway: This is largely a matter for urban planners, who must establish 'green waste' recycling systems that recycle nutrients into urban food production in every city and prohibit the dumping of food or green waste in landfills. Waste from agribusiness food chains must also be recycled.

6. *Reshape the world diet from one that degrades the planet and our personal health to one which protects and preserves both.*

Pathway: This is achievable through better consumer health education via schools, health care and medical services, the food industry (see point 2), media and social media, farmers' markets, etc. A transition from a 'European' diet (i.e. red meat, grains, and processed foods) to an 'Asian' diet (i.e. more fresh fish and fresh fruits and vegetables) will lead to a major reduction in the environmental damage caused by industrial food production, a large improvement in human health generally (along with reduced health care costs), lower greenhouse gas emissions, more affordable food, and a more varied and interesting cuisine.

7. *Reinvest 20 per cent of world defence spending in 'peace through food'.*

Pathway: Diverting a fifth of military spending to sustainable food for all will reduce the threat of war by around two-thirds,[6] as well as providing an adequate diet for all which is far more sustainable than the present one. The funds can be invested in developing urban food systems, deep-ocean farming, a Stewards of the Earth programme to restore the world's damaged landscapes and lock up carbon, and research into novel foods and production systems that tread more lightly on the planet. Renewable food will help to stabilise developing countries in particular, end hunger, and reduce the risk of civil and inter-state conflicts.

8. *Establish a global Stewards of the Earth programme consisting of hundreds of millions of former farmers and indigenous people.* They will work together to repair and replant the Earth's forests, grasslands, waterways, and wild landscapes over an area half of that which is presently cleared and degraded by unsustainable farming, mining, and development.

Pathway: This is a global initiative to repair ruined landscapes, replant forests and grasslands, restore natural waterways and wetlands, lock up carbon, and regenerate nature. Its workforce will consist primarily of willing farmers, former farmers, and indigenous people, who have a deep understanding of their own local environments and who are paid and funded through global levies on (a) military budgets and (b) food and natural goods, to repair the Earth.

With corporate agribusiness currently throwing hundreds of millions of small farmers off their land and replacing them with unsustainable industrial food systems, the need to retain their skills, knowledge, and love of country for restoration of the planet is urgent. A Stewards of the Earth programme will create a skilled workforce, on the spot, for environmental repair and restoration, prevention of extinction, and prevention of new pandemic diseases as global food production moves progressively into cities and the deep oceans. Its aim is to restore and rewild the global natural environment so that it can sustain all of our civilisation together with the Earth's existing biodiversity far into the future.

9. *Introduce a 'Year of Food' in every junior school on the planet, teaching respect, awareness, and appreciation of food.* Empower children to educate their parents about sustainable, healthy food and care for the environment that supplies it.

Pathway: As described in point 4, the information to do this already exists. Numerous programmes in urban schools are presently reintroducing how to grow and prepare simple foods. It can be done without altering the curriculum, but merely building a food theme into every subject taught – from science and maths, to languages, geography, social studies, and sport.

10. *Ensure the availability of family planning, education, and health care for women in all societies worldwide* (see Chapter 9).

Pathway: Voluntary population reduction is ultimately key to sustaining the world food supply and ensuring there is enough for all. Family planning needs to expand much faster and be coupled with equal opportunities for women and the acceptance of women as leaders in all walks of life.

11. *Establish a universal internet-based system for sharing food and nutritional and agricultural knowledge with all farmers and consumers.*

Pathway: Various organisations, public and private, are already working on this, but they need to collaborate better and more urgently. A globally funded programme and database would help. The world needs a virtual 'Library of Alexandria' of food.

What You Can Do

- Choose your food with great care. Understand which foods are produced sustainably – and which ones destroy the planet. Seek the former, avoid the latter.
- Use your consumer spending power to tell food retailers, supermarket chains, and producers that you highly value safe, healthy, sustainable, fresh food.
- Be happy to reward farmers fairly for looking after the landscape as well as producing safe, healthy, delicious produce.
- Eat fresh, eat local, eat sustainably, eat healthily. Learn what these things entail and how to distinguish between evidence-based food facts and claims driven

by unfounded beliefs or big industrial companies
seeking to turn food into profit.

- Eat less meat and more fresh vegetables and fruits. This
  will help end the clearing of land, the felling of forests,
  and the needless killing of wildlife. It will start fewer
  pandemics, emit far less greenhouse gas and toxic
  pollution, and save the Earth's life-giving resources, soil,
  water, and vegetation to support both humans and
  all other living creatures into the future. It will also
  improve your own health as well as public health in
  general.
- Consciously avoid highly processed 'industrial foods'
  and fast food that is unhealthy, contains too many
  chemicals and additives, is packaged in plastics and
  other toxic wrappers and which disrespects farmers by
  paying them badly. Eat with a clear eye to lifelong good
  health for yourself and your planet.
- Support 'one child fewer', family planning, education
  for women, and other voluntary means of reducing the
  human population and its impact on Earth's life
  support systems.
- Teach your children to respect and value food, as every
  previous human generation used to do until the
  modern age.
- Understand that our attitudes and values with respect
  to food will define the human future, for good or ill. If
  we choose food designed mainly for industry profit, it
  will ruin our health and wreck the Earth's ability to
  supply nourishment.
- Grow more of your own fresh food, and prepare and
  take pleasure in it. Make sure your garden soil is clean
  and free of contaminants, either by having a soil test
  or buying it from a reputable supplier. Learn how to
  control pests without the use of chemicals. Learn how
  to build fertile soils without using chemical
  fertilisers.

- Talk about the importance of a sustainable food supply with your family and friends. Build enthusiasm and interest in growing their own, eating fresh and naturally. Help to spread the word and share the good news that by eating more wisely we can heal and protect ourselves and the Earth that supports us.

- Set up a local food exchange with your neighbours and in your street. Every food garden has surpluses which can be exchanged with others for foods you do not produce, face-to-face or online. Look for food exchange apps that enable this in your locality. Organise local composting of garden, vegetable, and kitchen waste if you lack the space to do it at home.

- There is a growing array of new cookery and food books which can help cooks, foodies, and families to choose a sustainable, healthy diet. Recent examples are:
  - Dana Hunnes, *Recipe for Survival*, Cambridge University Press, 2022
  - Mark Bittman, *The Food Matters Cookbook*, Simon & Shuster, 2010
  - Sue Radd, *Food as Medicine*, Signs Publishing, 2016
  - Heather Wolfe and Jaynie McCloskey, *Sustainable Kitchen*, MennoMedia, 2020
  - Melissa Hemsley, *Eat Green*, Ebury Press, 2020
  - Michael Pollan, *The Omnivore's Dilemma*, Penguin, 2006.
  - Frances Moore Lappé, *Diet for a Small Planet*, Random House, 1971.

- Never forget, the future of food is the future of humanity. Good, renewable food is the key to our persistence on the planet.

9 'ONE CHILD FEWER'

The Problem

In barely the space of a single lifetime the human population has swollen from 2 billion to 8 billion and continues to burgeon at a rate of around 80 million (1 per cent) a year. If such growth were to continue unabated, there would be 21 billion humans on Earth by 2120. Anyone who considers the matter soon realises that the resources needed to support such gigantic numbers will run out – and there will be a painful population crash.

Overpopulation drives and compounds all the other threats which constitute our existential emergency. A large and growing human mass increases our impact on the Earth's climate; on the volume of poisons we emit into the biosphere; on rates of wildlife extinction, land clearing, and ecological breakdown; on the dwindling of key resources like clean water, soil, fish, and forests; on spreading food insecurity; on the risk of fresh pandemics and new wars; and on the number of people who will suffer and die or be forced to flee their homes to live in other lands due to any of these causes. While not deadly in itself, population overgrowth is the mainspring of the existential emergency that now faces every person on the planet. *None of the catastrophic threats will abate significantly while human numbers remain so high.*

The most eye-catching feature of the human population graph (Figure 9.1) in the last 10,000 years – that is, since we adopted agriculture and began to live in cities – is the

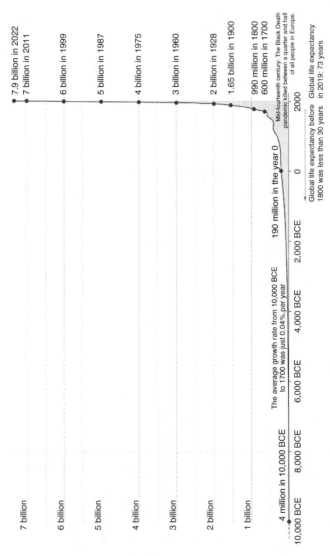

Figure 9.1 Human population growth 10,000 BP to 2021. Credit: Roser M, Ritchie H, & Ortiz-Ospina E. World population growth, 2013. Our World In Data. https://ourworldindata.org/world-population-growth. Based on estimates by the History Database of the Global Environment (HYDE) and the United Nations.

explosive growth that has occurred in the last three generations, when our numbers blew out fourfold. When such population explosions occur in nature, and the animal runs out of living resources, there is usually a mass die-off and its numbers crash abruptly back to where it started, or even below. This is shown by the famous case of St Matthew's Island, Alaska, where 29 reindeer were introduced in 1944 by the US Coastguard to help supply food to a small military outpost.[1] Unsupervised, the reindeer numbers boomed to over 6,000 by 1963 and proceeded to eat up all the available lichen (their main food source) and vegetation. When this point was reached, starvation set in and reindeer numbers crashed by 99 per cent in a single hard winter, leaving only 42 animals alive in 1966. By 1980 these had died out altogether, leaving the island with no large animals. The story illustrates the common principle in nature that when a creature overshoots its resources and climate takes a hand, its numbers can crash catastrophically, with unimaginable suffering for all.

That humans are subject to the same rule of nature is illustrated by the case of Easter Island (Rapa Nui), where scientists now consider that climatic changes reduced the ability of the deforested but well-populated island to feed its inhabitants, resulting in a series of steep declines. Professor Mauricio Lima of Santiago University, Chile said: 'The population on Rapa Nui lived – and live – on a small and remote island with limited resources, and we ourselves are living on a small and remote planet with limited resources. One of the lessons from this study is the importance of interactions between climate change, human population size and changes in the ecosystem.'[2]

Vast climate-driven famines in countries such as China and India and the continent of Africa over recent centuries show that the largest societies are not immune to such events. Now, in a globalised economy, food system failure almost anywhere impacts populations worldwide through regional food scarcities, mass migration, and skyrocketing world food prices.

The central issue of human population growth is not whether it is good or bad, but whether we can avoid a devastating crash caused by our outrunning the planet's ability to support us. Voluntary population reduction is therefore about sparing *billions of people* needless and agonising death by starvation, war, and disease, which will be the result of a collapse in our resources. While most people hope this may never happen or, if it does, that we can do something to minimise the damage, the underlying point is that a crash is the *inevitable* consequence of uncontrolled population growth. To prevent the crash, we have to halt and reverse the growth. There is no real alternative.

Of course, this is not popular with some people. Many, including would-be parents, politicians, religious leaders, and business leaders, want us to have more babies in order to 'grow the economy', increase a nation's relative significance, outpopulate some competing group, belief, or culture or simply assert what we deem as our rights (at the expense of the rights of *everybody*). Consequently, there is a widespread and growing global attempt to silence discussion both of overpopulation and of ways to prevent it.

Often too blinkered to realise it, people who urge a larger population are in fact calling for:

- increased scarcity of vital resources such as water, soil, fish, and certain minerals, leading to a higher risk of wars breaking out over these diminished resources, as well as growing floods of refugees and displaced people;
- accelerated climate change caused by increased use of fossil fuels and land clearing;
- widening pollution, environmental degradation, and extinction of species, leading to a reduction in the Earth's ability to sustain human life;

- higher food prices for all, rising insecurity in the global food supply, and greater risk of famine;
- more child deaths and suffering;
- increased risk of pandemic diseases, poorer levels of population health;
- an increase in mass population movements, with world migrant numbers potentially reaching 1 billion a year;
- increased risk of megacity and government failure;
- increased risk of worldwide economic and civilisational collapse;
- housing, food, and other basic goods that are unaffordable to the young or the poor.

Why any sane person would want such outcomes for themselves or their children is not easily explained. The only answer is that people may choose to ignore the consequences of their personal decision from selfish motives, or else are gambling that they can somehow avoid them – while everyone else suffers.[3]

What Is Overpopulation?

There is much confusion, often deliberately fomented by vested interests, over the term 'overpopulation' as it applies to humans. Put simply, if any of the Earth's vital indicators – such as water, soil, forests, ocean health, air quality, climate, or the environment – are deteriorating as a result of human activity, then there are too many people, and the Earth is overpopulated in terms of the numbers it can safely support in the long run.[4] Overpopulation is not a matter of personal or group opinion: it is defined by measuring human impact on the planet's physical resources. If they are improving, then our population is in balance. If they are deteriorating, then we are headed for trouble.

In short, the more the population grows, the less land, water, clean air, forests, fish, minerals, and other vital resources there are to go round on a finite planet. Civilisation compensates for the lack of resources by pushing up the monetary prices of goods that are in short supply, leading to a situation, for example, where young people cannot afford to own their own homes and many cannot afford adequate diets, water, shelter, health care, and education to meet their needs. The result of this deprivation is increased anger, unrest, crime, conflict, and civil war – with the poorest nations being worst affected. That is the immediate price of overpopulation.

The 2020 world population increase of 81 million people was the highest in history, even though the number of births per woman continued a modest decline.[5] This is because the number of parents on the planet is still increasing – even though many are now opting to have smaller families. Demographic projections point to a modest decline in the rate of growth to 2050, but the overall world population will then probably still be growing by around 50 million people a year, to a total of around 10 billion.

In 2021, a record 84 million people were forced to leave their homes due to famine, war, or political oppression.[6] Among these, climate refugees alone were estimated at 21.5 million. In addition to these, voluntary migrants – many leaving unsafe areas to seek safer lives in developed countries – totalled 272 million (before the Covid-19 pandemic temporarily restricted world travel).[7] Thus, in a typical year, more than a third of a billion people – 4–5 per cent of humanity – seek refuge and a better life outside their own country, consequently exerting growing pressure on the borders, major cities, resources, and infrastructure of other, more stable, nations.

However, with climate, resources, and political stability deteriorating more rapidly as the world population swells, the numbers of global migrants and refugees are expected to climb sharply. The World Bank, for example, projects a tenfold increase in climate migration by 2050.[8] The UN cautions that total international immigrant numbers could, under some circumstances, reach 1 billion by 2050.[9]

Gagging Discussion

One of the greatest myths surrounding overpopulation is that it creates economic growth. Even a casual survey of the world reveals that natural population growth is lowest in rich countries – and highest in the poorest. It is countries that are quickest to reduce their birth rates that achieve higher living standards for their people fastest – not the other way around. Calls for increased population to promote economic growth are both untrue and calculated to serve the interests of particular groups (usually those who want cheap labour or who sell real estate) at the expense of everyone. The economic, social, health, security, and environmental benefits of lowering population and of better family planning[10] are being widely suppressed or shouted down in public discourse.

Population expert Dr Jane O'Sullivan warns that 'A dangerous taboo has developed over global discussion of population growth.'[11] She continues:

What this issue is really about is trying to prevent a huge die-off of people. Overrunning our natural resources can only lead to more deaths from starvation, conflict and disease – because we are already in overshoot. The only alternative pathway is voluntary restraint on the number of human births while, at the same time, we try simultaneously to reduce our consumption and environmental impact per person. To choose not to reduce population is to choose human suffering on a vast scale.[12]

O'Sullivan cautions that the longer we stay in overshoot, the further our population has to fall before we can achieve sustainability, so time is ticking away while we argue over the issue. 'Fertility is falling much more slowly since discussing overpopulation became politically taboo. This means we are more likely to see a population crash', she says. 'How can we solve the population problem if we are not allowed to name it?'[13]

Family Planning

The single best way to reduce overpopulation voluntarily is through family planning – and by individuals taking a voluntary decision to have fewer babies, or even none at all. Providing women with education, health care, and increased economic opportunities also leads to a lower birth rate and to healthy children being raised in better conditions. Access to safe, voluntary family planning is a universal human right. It is a key factor in reducing poverty.[14]

The benefits of increased family planning include the following:

- It allows countries and families to invest more in each person, accelerating economic growth and social equity.
- It is the single, best way to raise the living standards of the poor.
- It leads to fewer child deaths.
- It slows and eventually reverses the scarcity of land, water, and other resources, leading to more for everyone.
- It is one of the most effective ways to reduce global carbon emissions.
- It lowers the human impact on the living environment, reducing pollution, deforestation, overfishing, and other ecosystem destruction.
- It saves the lives of mothers and infants, improves health care across society, empowers women, and frees them for leadership roles.

- It is much safer for women than black-market contraceptives or abortions.
- It is voluntary, costs little, and can easily be offered free of charge to all.

Dr Jane O'Sullivan contends that the commonly voiced claim that population growth is 'good for the economy' is nothing more than a cruel hoax.[15] The evidence shows conclusively that developing countries with low and falling birth rates have the strongest rates of economic growth – while those with rising populations struggle to escape the poverty trap. The claim about population growth is usually promoted by vested interests who have the most to gain from creating scarcity, such as land developers. It is a common misapprehension in business that more people means more customers, when in fact having fewer, more affluent customers who spend more makes a business far more profitable. Strong population growth, on the other hand, means lower wages for workers and so reduced spending by consumers, which makes businesses less profitable.

Most importantly, effective family planning worldwide will temper and abate all 10 of the catastrophic risks – including the risk of nuclear war over dwindling resources. For this reason, budgets currently devoted to weapons would achieve more for world peace if part of them were assigned to family planning instead of to killer robots, for example. Family planning needs for the entire world could be met for less than a third of the cost of a nuclear submarine.

With increased voluntary family planning worldwide, there is no need at all for compulsory or coercive population control – and this is now recognised universally. Yet still the groups who place their own selfish needs (economic, political, and religious) above those of humanity try to silence discussion and gag the agencies who seek more attention and resources for birth control. They are advocates of disaster and agents of catastrophe.

To achieve a sustainable population, the global birth rate needs to fall to less than two babies per couple. Even then, it will take many decades before the population begins to contract because the number of couples is still growing and people are living longer. The fairest way to achieve a smaller human population quickly is 'one child fewer', in which families voluntarily choose to have one child fewer than planned.[16] Thus, a two-child family becomes a one-child family, a five-child family becomes a four-child family, a one-child family becomes a no-child family, and so on. If adopted widely, 'one child fewer' could help reduce the human population voluntarily by 3.5 billion by 2100, the UN considers.

Worldwide, there is a growing trend among younger people to choose to be 'child free' – to have no children at all during their lifetime.[17] While it defies many cultural traditions, the trend to have no children at all is growing in all countries, driven, according to young people themselves, by not wishing to expose a child to the catastrophes they foresee as human populations start to crash. In one US study, as many as 1 in 4 adults said they preferred not to have children.[18] It found that child-free people were just as happy and came from the same social backgrounds as people with children. The evidence suggests that more and more women worldwide are redefining their traditional role as mothers in favour of roles as leaders, breadwinners, and career professionals and are comfortable with this – although they still face stigmatisation from others who are, perhaps, parents or who hold strong religious views. The decision to be child-free takes courage, wisdom, and clear sight.

What We Can Do

1. Increase global and national funding for voluntary family planning, especially in countries with high birth rates. It will cost only an extra US$340 million a year to

deliver family planning to every woman on earth who wants or needs it.

2. Divert military spending to family planning to reduce the threat of war caused by overpopulation.
3. Launch a universal education campaign to explain to all citizens the dangers of overpopulation and the economic, health, and other benefits of choosing to have smaller families.
4. Provide support and advice for all individuals and families who wish to be child-free or to have small families.
5. Eliminate government subsidies and support for higher birth rates.
6. Provide legal and other forms of protection against population extremists.
7. Elect or appoint wise women to leadership positions in all walks of life, including government and civil institutions; industry; business and corporations; political groups and parties; religious institutions; community bodies; international agencies; clubs; and service organisations.

What You Can Do

- Have one child fewer than planned.
- Aim for two children or fewer per couple.
- Do not be afraid to advocate for a lower birth rate and family planning, as you are also advocating for far less human suffering and needless death, including the deaths of children.
- Celebrate depopulation and recognise its good outcomes and many benefits for all.
- Recognise that a smaller population is the only way that human civilisation can be sustained on this planet – this is an act of great care for future generations.

10 HEALING TECHNOLOGICAL MAYHEM

The Problem

Every catastrophic and existential threat which humans face today is the direct result of our use, overuse, or misuse of technology. Technical advances, while wonderful and much admired when they first appear, often contain the seeds of disaster if applied in ill-considered ways.

Two simple examples will suffice. Coal was a boon when it was first used to power industry, railways, and shipping, heat homes, and generate electricity. But as human numbers rose into the billions and our demand for energy grew, coal became a source of deadly air and water pollution and a menace to human existence as the chief driver of global heating. Thus, it was not coal that was at fault, but the thoughtless and prodigal ways humans used it, which turned the blessing into a curse.

Second, agriculture was a boon when it enabled humans to settle in towns and cities and acquire new skills instead of having to spend their lives finding food. It is the platform that launched the modern technological age and from which megacities were built. From the 1970s, however, the Green Revolution unlocked the farming potential of the planet, which in turn lowered child mortality and helped to trigger the human population explosion and today's universal overclearing and destruction of forests, grasslands, soils, climate, water, and wildlife. Thus, a benign technology, through overexpansion, has become a menace to the human future. Likewise, public health measures – vaccines and antibiotics – while saving countless lives, have had the knock-on effect of causing populations to explode.

The march of technology gives rise to the notion of the 'vulnerable world'[1] in which technology both benefits us and, at the same time, threatens to inflict catastrophic harm. As Professor Nick Bostrom of Oxford University's Future of Humanity Institute explains:

Scientific and technological progress might change people's capabilities or incentives in ways that would destabilize civilization. For example, advances in DIY biohacking tools might make it easy for anybody with basic training in biology to kill millions; novel military technologies could trigger arms races in which whoever strikes first has a decisive advantage; or some economically advantageous process may be invented that produces disastrous negative global externalities that are hard to regulate.[2]

And Dr Brian Green, director of technology ethics at Santa Clara University in the United States, warns: 'Numerous emerging technologies promise great, but finite, benefits while also placing humanity at tremendous risk, even the risk of extinction.'[3] The main concern is that novel technologies with the potential for both good and evil uses and outcomes are being developed by researchers and corporations worldwide at a phenomenal rate, often with little regard for ethics or the disasters they may unleash.

Technology has destroyed civilisations in the past: many native peoples of North and South America, Australasia, Africa, and Asia fell victim to European military, industrial, and agricultural technologies. More will do so in the future. What is new is the potential of today's ultra-powerful technologies – specifically biotechnology, nanotechnology, artificial intelligence, robot killers, and mass surveillance – to bring down the whole of humanity if their use is not carefully overseen and regulated globally, by governments, and by consumers. The following sections briefly describe the threat presented by six fields of ultra-science.

Biotechnology

Advances in biological science in the twenty-first century have greatly accelerated our ability to manipulate and reprogramme bacteria, viruses, and other living organisms, opening up an array of potential new catastrophic threats in the form of novel plagues, bioweapons, and bioterrorism.[4] In contrast to nuclear weapons, biology is cheap to perform and relatively easy to do, given the right knowledge, making it more accessible to malignant actors. Of particular concern is the development of synthetic biology,[5] which involves the deliberate re-engineering of wild viruses, plants, and other organisms, mostly to make them more useful – but sometimes, more deadly. Because it is cheap and offers big commercial rewards, synthetic biology can be carried out almost anywhere, in insecure laboratories, by unskilled staff, and without ethical approval or government oversight. Consequently, the risks of laboratory accidents, escapes, and malicious release are high. Examples of the dangerous misuse of biotechnology include the anthrax attack in the United States[6] and the artificial creation of a killer strain of influenza.[7] Concerns persist that the Covid-19 pandemic may have originated with the escape of an artificially engineered virus.[8]

Nanotechnology

Nanotechnology is the manipulation of materials at the 'nano' (billionths of a metre) scale to endow them with novel properties and improve their durability and performance. This means nano-substances are only a few atoms wide and can penetrate into many places where they are not wanted. The chief danger of these products is from 'nano-pollution', the distribution of ultra-fine particles in air, water, the food chain, and the oceans and their ability to infiltrate all life on Earth, including the blood–brain barrier or the mother–embryo barrier, potentially injuring both brain and baby.[9] A major issue with these very small substances is that once they have been released into the environment they can never be recalled and, if dangerous, can theoretically continue to cause harm forever. In this sense nano-pollution and nano-poisoning pose a catastrophic risk to human life which is poorly understood, has never been properly evaluated, and is unregulated and uncontrolled in terms of scale or impact. Without being asked, all human beings have become the unwitting guinea pigs in a worldwide, potentially deadly, nanochemistry experiment.[10]

Hailed as 'a new Industrial Revolution', molecular self-assembly is another recognised risk.[11] This involves self-copying by minute devices, substances, or liquids, which could in theory escape human control and propagate forever or until they run out of the necessary resources. Self-assembly concerns range from the idea of robots reproducing themselves regardless of what humans may want, to 'grey goo', a self-propagating liquid. Scientists in the field tend to make light of these dangers, but have not proved they are impossible nor attempted to limit their development.

In nanotechnology, as in all the ultra-technologies, there is a global absence of ethics, independent oversight, or government regulation of both experiments and the development and release of commercial products. As with coal and chemistry, science and industry are pursuing the rewards while turning a blind eye to the risks.

Artificial Intelligence

Artificial intelligence (AI) is the teaching of machines to 'think', or at least respond in a human-like way for themselves, with minimal input from actual live human beings. In the 1940s, science fiction author Isaac Asimov articulated the fear and concern this engenders. In 2014, British physicist Professor Stephen Hawking warned, 'The development of full artificial intelligence could spell the end of the human race.'[12] And in 2021, high-tech billionaire and head Elon Musk stated, 'AI is far more dangerous than nukes ... it scares the hell out of me.'[13] Despite such warnings, the race to empower machines to act autonomously has accelerated and is now widely regarded as out of control. Safety engineer Bela Liptak summarises: 'Today, the AI development process is uncontrolled, the developers have different goals and some of these goals are undesirable or even dangerous.'[14]

Popular fears about AI tend largely to focus on the capacity of machines to replace humans in basic jobs – a process that has been going on since the industrial revolution but which is now accelerating as machines learn to wield higher-order skill sets such as those of doctors, writers, or lawyers. Job automation, for all its advantages, is now a threat to many humans and promises to widen socio-economic inequality between the tech-haves and the tech have-nots.

What bothers the experts more, however, is computers reaching the point where they can reprogramme themselves and so set goals which do not align with human needs and, indeed, may well conflict with them – such as copying themselves ad infinitum. This could unleash a phase of technological development over which humans have absolutely no control.

Other applications of AI that arouse powerful concerns include 'killer robots', universal surveillance of individuals, global loss of security and privacy, the undermining of politics via the widespread dissemination of fake information (see Chapter 11), instability in world financial markets caused by investment algorithms overreacting to trends, and the use of AI to make 'designer' diseases or dangerous novel materials far removed from those that exist in nature.

Robot Killing Machines

Machines that can kill people without requiring a human command to do so are being deployed by leading military powers worldwide, including China, Israel, South Korea, Russia, the UK, and the United States.[15] Researchers now envision a future where entire armies may consist of self-controlled robotic devices instead of soldiers,[16] with terrible consequences for any civilian populations upon which they may set themselves loose. Unfortunately, there is now a global robot arms race.[17] To cement their technical hegemony, the same nations are also blocking efforts by other countries to prohibit autonomous weapons systems through a new international treaty.[18]

Robot killing machines are a branch of AI – deadly devices empowered to choose their own targets and destroy them. However, the risks of AI are greatly multiplied when coupled with nuclear or biological weapons of mass destruction, or swarms of drones armed with powerful explosive missiles. They pose a threat both in the enormous power they confer on their makers to inflict mass death and destruction on those they hate or fear, and also in their potential to go wrong and attack human populations autonomously or start wars by mistake. Because of their enormous, mindless destructive power, robot killing machines warrant special priority for an international ban or restrictions, and their development demands global civilian control and oversight.

Universal Surveillance

The advent of quantum and other ultra-powerful computers, with vast speed and colossal data storage and
retrieval capacity, is opening the way for governments,
corporations, and other players (including criminals) to
spy on literally everyone from birth to death, thus eliminating all human freedoms forever.

This threatens to usher in an era in which the deeds,
words, personality, and even the thoughts of every individual are subject to automated scrutiny every moment of
their lives. Through the mining of this data, individuals
can be controlled – whether in their spending decisions,
their political or religious views, their personal opinions,
their mindset, or their actions – simply by threatening to
expose all that surveillance knows about them, which is
often far more than they know about themselves.

Already most citizens are under scrutiny by CCTV cam-
eras and facial and other forms of identification; all
our financial transactions, our internet and smartphone
actions, as well as our health and other private records
are readily available to those with the computing power
to access them, whether or not they have permission. This
gives rise to the fearful possibility of a de facto global police
state in which every individual can be controlled, silenced,
or forced into obedience. This has dire implications for
humanity at the most dangerous period of our existence.

A brief reflection on history will quickly reveal the
dangers of silencing, throughout their lives, all our
change agents, thinkers, leaders, radicals, inventers, and
innovators – everyone, in short, who stands out from the
herd. In such a society, people like Socrates, Jesus,
Mohammed, Gandhi, Galileo, Newton, da Vinci, Darwin,
Mandela, and Einstein would be quietly airbrushed out of
history because of the 'troubling' ideas they propounded
and the risks these posed to those controlling the status
quo. No radicals, no change. No change, no future.

In the modern era, such global censorship of thought, ideas, free speech, and originality will be used by the powerful to silence the warning voices who speak out on issues such as climate change, nuclear war, global pollution, extinction, and the other catastrophic threats. This is already happening with the jailing, in many countries, of young people who protest over climate and the environmental damage caused by corporations.[19]

The evolution of a global police state can also be used to end or to limit individual freedoms and control behaviours – except for those approved by the controllers.[20] Such an obliteration of society's ability to question those in power overturns the principles of democracy and leads directly to the end of civilisation – because nothing can be done to prevent it. The censorship and suppression of scientists, academics, doctors, reformers, young protesters, activists, and environmentalists, in particular, eliminates the voices of those who warn that the human enterprise has passed the safe boundaries of its existence. In short, universal surveillance can be used to prevent humans from saving ourselves, even if as individuals we clearly understand the need to act.

Unknown Boundaries

In the same way that the awesome power locked into a single atom of uranium was revealed to the world with the birth of the nuclear age in the 1940s, some people remain concerned that there are as-yet-unknown physical boundaries and phenomena which humans may accidentally breach or release through our unquenchable appetite for new knowledge and discovery. An example is the fear that instruments such as the Large Hadron Collider, in smashing together the elementary particles which comprise the universe, may inadvertently give rise to micro-black holes, dangerous exotic particles, or trigger subatomic chain reactions that threaten the existence of

matter. Particle physicists dismiss these fears as fantasy[21] –
but the fact remains that civil scrutiny of such potentially
risky science is minimal. As in other ultra-sciences –
such as biotechnology, nanotechnology, and information
technology – those with a vested interest in the research
are also in charge of its ethics and oversight and the public
is denied a say about matters that affect their future.
A world-ending risk, however slight the experts may
deem it to be, deserves to be treated with due caution and
the whole of society should be consulted.

The lack of progress in addressing any of the major
global risks to civilisation and human existence is graphic-
ally chronicled in the World Economic Forum's 2022
Global Risks Report.[22]

What We Can Do

The core issue with the ultra-technologies described in
this chapter is the absence of civil or ethical oversight by
society and the lack of any real ability on the part of
governments, global agencies, or humanity at large to
check, rein in, or even query research that may give rise
to new catastrophic threats.

Motivated chiefly by greed and/or fear, rather than concern for society, ultra-technology research is proceeding at
full pace worldwide and is now effectively out of humanity's control. A range of high-tech genies are being released
from various bottles, because of the advantages or profits
they appear to promise, without due consideration of their
disadvantages or dangers if badly, excessively, or mali-
ciously applied. Most of this work is secretive, taking
place in universities, weapons research establishments,
and commercial laboratories that are closed to public
view and often beyond the scrutiny even of governments.

Having seen the mess which uncontrolled use of coal,
chemicals, industrial food production, and plastics have

already made of our planet, it is unquestionably dangerous for humanity to incur similar risks of pollution, war, disease, and inequity arising from biotechnology, nanotechnology, AI, and the other ultra-sciences. While the dangers are increasingly recognised by the public as well as by some scientists, attempts to control them are largely confined to particular categories of technology, such as killer robots.

The issue which has so far escaped global attention is that *almost any advanced technology*, released into a world of 8–10 billion people all demanding new products and services on a resource-stressed planet, has the capacity for serious negative outcomes, whether it is deployed with good intentions or not. The human technology machine has run amok, not because the technology is necessarily bad but because its overuse and misuse are inevitable and its side-effects, such as pollution and mass poisoning, are usually ignored until it is too late to remedy them.

Instead of attempting to regulate each new technology and its particular applications one at a time – a fragmented endeavour doomed to fail (as it already has in the case of chemistry, see Chapter 6) – it makes far more sense to regulate *all* human technologies against their overuse, abuse, or misuse. Clearly, such oversight will have to occur at global scale as it did, for example, in the case of the Law of the Sea,[23] because no single country or group of countries can control the march of technology. Furthermore, no technology can be limited to a single country; history shows it invariably diffuses to others and then worldwide.

The central proposal advanced here is that there should be a Global Technology Convention, which provides for the oversight of technology development by civil society; a fair and objective analysis by an independent body of its impacts, good and bad, on society; and enforceable regulation to restrict or ban those technologies and applications deemed too dangerous or adverse to the human future.

Several countries used to have formal technology review processes – for example, the US Office of Technology Assessment – but discontinued them due to pressure from industry. The institutional knowledge in legislatures about how to regulate powerful new technologies has been lost as a consequence. This has resulted in a growing catastrophic threat to humanity from the unexpected outcomes of technology use, abuse, and overuse.[24]

Just as the Law of the Sea seeks to rein in overfishing, pollution, and conflict on the high seas outside national jurisdiction, as well as influencing behaviour within national zones, a Global Technology Convention can help rein in the deleterious use of technologies both globally and nationally. As with all such conventions, individual countries will be invited to participate, sign, and ratify it. An independent world scientific organisation parallel to the Intergovernmental Panel on Climate Change[25] should be established to monitor, document, and report on advanced technology developments and their potential risks. These risks could then be controlled and policed under an emerging body of international law, as is the case with climate and environmental law.[26]

The Solutions

The following is a list of essential actions at global and national scale if humanity is to avoid future technology-driven crises akin to the climate, chemical, and extinction crises that are now engulfing us.

1. The world must establish a Global Technology Convention as soon as possible, along with an Intergovernmental Agency for Human Technologies (IAHT) to assess risk and recommend action. The Convention could form a major part of an Earth System Treaty.

2. All research into advanced technologies and ultra-sciences should be subject to a strict code of ethics, overseen by independent representatives of civil society who have no vested interest in either the research or its outcomes. It should be a basic requirement for all science to conduct precautionary research into potential dangers and drawbacks of new technologies and report publicly on them.

3. It is time for all scientific disciplines to adopt a 'first do no harm'[27] code of ethics for their practitioners, as medical doctors do, to reduce the likelihood of their science being used for evil or dangerous purposes. This code should first be adopted voluntarily by universities and scientific professional bodies, but failing this it should be imposed by civil and international law.[28]

4. As a matter of urgency, new human rights[29] should be established in the areas most affected by the ultra-sciences:

   • a human right prohibiting the mass surveillance of populations;
   • a human right prohibiting the toxic pollution of human beings, their water and food supply, and the environment with nanoparticles and untested chemicals;
   • a human right prohibiting the engineering of dangerous biological entities which may give rise to future pandemics and unknown diseases;
   • a human right prohibiting the use of killer robots against civilian populations.

Pathway: While human rights in themselves do not prevent the evils they define, they represent a conspicuous standard for the whole of world society and they can create pressure on governments, corporations, and individuals to comply with them. They also create a legal avenue for those adversely affected by technologies to seek redress and a means of publicly shaming those who breach them. These human rights can raise awareness across the whole of society about the dangers of uncontrolled technology development.

What You Can Do

- As a voter, demand laws that publicly disclose advances in AI and biological and nanoscience, so that there can be free and fair public debate about which aspects of these powerful new technologies should be allowed and which should be restricted or banned.
- Demand new human rights to protect your rights and freedoms from the intrusive use of ultra-sciences and technologies by governments or the private sector.
- Take a moral stand against machines which can kill humans based on a decision made by AI or robotically. Protest their use by any nation or government.
- Take a moral stand against universal data collection, mass surveillance, and their misuse. Demand constitutional reform to protect your freedom and private data from spying by governments, other nations, criminal organisations, or commercial corporations.
- Support a Global Technology Convention to curb the risks of uncontrolled technological developments. Discuss the need with colleagues, family, and friends.
- As a consumer, avoid buying products or shares in companies whose technologies threaten to harm, exploit, and impoverish other people or damage the landscape, air, water, food supply, or other resources needed for human survival – or who spy on their customers. Do not reward the wealthy for selfish behaviour.
- Require ethics, decency, and fairness from all with whom you deal. Enforce them by your personal spending, investment, and voting choices.

11 ENDING THE AGE OF DECEIT

The Problem

Perhaps the deadliest pandemic ever to strike humanity is the plague of deliberate misinformation, mass delusion, and unfounded beliefs which is engulfing twenty-first-century human society.

Whether generated by the fossil fuels lobby, certain media or other corporate interests, the anti-vaccine lobby, religious fanatics, political and ideological extremists, well-meaning simpletons or nutcase conspiracists, a deluge of utter nonsense is rapidly inundating the human species. In the short run, it may appear irritating, even occasionally amusing – but in the long run it lays the ground for spreading public ignorance of the risks we face and the need to overcome them, the failure of governments to prevent disaster, and, ultimately, the collapse of human civilisation.

The authors of the Doomsday Clock's 2021 report warned: 'This wanton disregard for science and the large-scale embrace of conspiratorial nonsense – often driven by political figures and partisan media – [has] undermined the ability of responsible national and global leaders to protect the security of their citizens.'[1] The dissemination of lies was increasing the danger from established threats like nuclear weapons, climate change, and pandemic disease, they added.

In a study of the threat to human knowledge from lies, Jevin West and Carl Bergstrom declare, 'Misinformation has reached crisis proportions.'[2] They conclude:

It poses a risk to international peace,[3] interferes with democratic decision making,[4] endangers the well-being of the planet[5] and threatens public health.[6] Public support for policies to control the spread of severe acute respiratory syndrome coronavirus 2 (SARS-CoV-2) is being undercut by misinformation, leading to the World Health Organization's 'infodemic' declaration.[7] Ultimately, misinformation undermines collective sense making and collective action. We cannot solve problems of public health, social inequity, or climate change without also addressing the growing problem of misinformation.[8]

Disinformation – the deliberate spreading of misinformation – can be murder: statistics worldwide reveal that Covid-19 death rates were far higher among the unvaccinated – in the United States, where vaccine denial is pervasive, it is 13 times higher.[9] Some researchers have even described the flood of nonsense as a new form of warfare, declared by one part of humanity against the whole – including themselves – using the global internet. The scholar Herbert Lin claims, 'Cyber-enabled information warfare has also become an existential threat in its own right.'[10] And Dr Steven Novella, editor of the online journal *Science-Based Medicine*, comments:

It's also clear that social media has given psychopaths and con artists the keys to the kingdom. It now pays, big, with little upfront investment, to spend a lot of time and energy crafting and spreading misinformation online. And not just isolated bits of misinformation, but a web of distortions and lies that are woven into a psychologically compelling narrative.[11]

Indeed, production of misinformation has attained global industrial scale, with the fossil fuels industry establishing a worldwide campaign costing hundreds of millions of dollars to deceive the public and government over the dangers of climate change and the role of fossil fuels in causing it.[12] Through purpose-built 'lie factories', the US$7 trillion petroleum sector (coal, oil, gas, and petrochemicals) has sought to misinform, manipulate, and sabotage world efforts to rein in climate change by corrupting governments, distorting public discourse, and circulating falsehoods. Its methods, adopted from those of the tobacco industry, are detailed in a report by researchers from three universities (Harvard, Bristol, and George Mason)[13] and in the famous book *Merchants of Doubt* by Naomi Oreskes and Erik M. Conway.[14] A study by Oxford University revealed evidence of formally organised social media manipulation campaigns in 48 countries,[15] describing them as 'a critical threat to public life' that was now 'big business'.[16]

In a *Scientific American* article calling for a global effort to combat the spread of fake science and deliberate misinformation about scientific findings, Dr Peter Hotez warned: 'Antiscience has emerged as a dominant and highly lethal force, and one that threatens global security, as much as do terrorism and nuclear proliferation.'[17] Indeed, people who spread lies are the world's new terrorists, causing far more harm, death, and destruction than their predecessors in the 'War on Terror'.

The complicity of the media – world as well as national and local; traditional as well as social media – in disseminating false information under the pretext of 'balanced reporting' – is a core part of the modern phenomenon. The lie factories cannot flourish without compliant messengers to carry their falsehoods.[18] Some media have even made the spreading of falsehood a core part of their business model, gambling that a big enough lie will attract 'more eyeballs' to their TV and internet sites, which the corporation then converts into advertising revenue from the corporate sector. Thus, the media have made the remarkable discovery that misinformation is more profitable than telling the truth. In time, such organisations end up cultivating loyal audiences who either love conspiracy or are just more credulous and ill-educated than most people regarding what they are told.[19]

However, 'fake news' is just one aspect of a very much larger human system of belief in things that do not truly exist. In my book *Surviving the 21st Century*, I identified the four chief human delusions – money, politics, religion, and the human narrative (stories we tell about ourselves). Money, clearly, has no existence outside the human mind – yet it drives our society and our overexploitation of the Earth. Politics, though vastly entertaining, often promises far more than it can deliver and lies are spread about opponents' promises. All religious beliefs are contradicted by every other religion, so invalidating them all. And in the human narrative, we prefer a 'Hollywood' image of humanity as invulnerable superheroes, which does not exist outside the realms of fiction or mass entertainment. Each of these beliefs is dangerous in its own way – money, because it drives the blind destruction of a habitable planet; politics and religion, because they prevent us from solving our problems through the social divisions they breed and the social unity and co-operation they undermine; and the human narrative because it encourages us to imagine we are some kind of super-species that can transcend all its problems without doing anything to actually understand or solve them.

For around 300 years, science has sought to dilute the human tendency to delusion by presenting an objective, tested view of the world and how it works – and constantly retesting it to make sure it is correct. No other belief takes such an approach. That groundwork of fact and understanding ushered in by the Enlightenment gave rise to the greatest achievements of civilisation of all time. Combined with money, politics, and religion, science has built the modern world. Now, in a new Dark Age, fantasy, superstition, delusion and denial are in the ascendant once again – with the deep irony that the great catastrophic risks are all founded on our misuse of science, mostly for monetary or political ends.

The motives behind the spreading of lies and conspiracies are many, but principal among them are monetary greed, political advantage, malice, hatred of 'others', and ruinous ignorance. Russia, for example, stands accused of using misinformation to sabotage American politics,[20] while the US Republican Party appears to have embraced misinformation because its voters are so susceptible to it.[21] Those who spread falsehoods are seldom aware they will be destroyed along with the rest of human civilisation – or else manage to fool themselves that this will not be the case. In a few cases, nihilistic individuals may actually wish to wreck civilisation and use the dissemination of falsehoods to set it against itself.

The problem of mass delusion is compounded by signs that humans today are less intelligent than they were a generation or two ago. Recent research has found that human IQ has declined by around 13 points since the mid-1970s.[22] While the cause is still uncertain, evidence is mounting that the fall in human intelligence coincides with the increase of nerve poisons in the living environment and with other observed brain ailments, especially in the young. What this indicates is that humans, due to their chemically injured brains, may now be more easily deceived and more susceptible to misinformation than they were a generation or two ago.

If humans are losing the capacity to reason, or have become addicted to fantasy, then there is no saving them in an existential emergency – because so many will not accept it as real or do enough to prevent it. Indeed, they may actively seek to undermine attempts by the rest of humanity to save itself – as in the case of the army of propagandists and 'useful idiots' employed by the fossil fuels sector to spread their untruths on social media.

Global misinformation may seem a modest threat in comparison with, say, all-out nuclear war, but it is no less deadly
in the long run because it disables the very quality on which
humans most pride themselves: the ability to think, understand, and act rationally to preserve our species.

The Solutions

1. *Reframe all our economic, political, religious, and narrative
discourses to place the survival of civilisation and the human species
as the common shared goal for all.* Call on everyone to participate in developing a realistic global plan for survival.

Pathway: This requires worldwide action by leaders at all
levels, the media, teachers, religious leaders, and concerned
citizens. The best way to achieve a global plan of action is
through an Earth System Treaty, negotiated internationally
by all countries, and signed and supported by them as
a declaration of intent to fix our broken planet. The evidence in this book underpins the arguments they can use.

2. *Dematerialise the world economy to prevent the overuse, pollution, and destruction of limited resources such as air, water, soil,
forests, and wildlife.*

Pathway: This has been outlined by the UNEP's
International Resource Panel. It should be coupled with
new thinking about the circular economy, industrial ecology, green manufacturing, recycling, and growth of the
creative economy – and subsidised, if need be, by governments so as to transfer jobs from the 'old' to the 'new'
economies as smoothly as possible.

3. *Encourage the world's religions to take more courageous and
informed moral leadership over issues that threaten humanity as
a whole, and to set aside their differences for the common good.*

Pathway: Religious leaders of all denominations must
embrace the future and the challenges it presents, rather

than clinging to the past – or risk being seen by society as redundant. They must focus on the evidence and what it says about our situation, as well as the moral good.

4. *Through citizen pressure, refocus world, national, and local politics on human survival.*

Pathway: Politicians are gradually learning that the path to political survival lies in attending to the concerns of their voters about *human survival.* However, all politicians need to undergo a short education in existential risk, as many still discount it, or are ignorant or lack good information about it.

5. *Inspire a new narrative about humans which prizes co-operation, tolerance, restoration, cleansing, and sustaining over older, more divisive, selfish, and destructive values.*

Pathway: In a way, this revisits the more co-operative, altruistic movements and belief systems of the past, rather than the savage legacy of the twentieth century. People are ready for a happier, less murderous, and less exploitative human story – and it is the great opportunity of storytellers to tell it.

6. *Establish a World Truth Commission (WTC) with power to investigate the veracity of claims made by prominent individuals or organisations that attempt to deceive the public, and report its findings openly.*

Pathway: Established on a similar basis to the International Criminal Court, the WTC will be a tribunal with expert scientific and legal staff empowered to investigate any claims by prominent bodies or people that are likely to adversely affect the catastrophic risks facing humanity. It will have the power of exposure in the same way that Amnesty International and Human Rights Watch monitor, name, and shame persistent offenders.

7. *Establish a World Integrity Service (WIS), to provide independent validation of the integrity, truthfulness, ethics, and honesty of organisations and corporations.*

Pathway: Just as various professions – for example, law, medicine, engineering, and chartered accountancy – stand behind the professional skills and integrity of their professionals, there is a need for a world body to validate the integrity and truthfulness of organisations, their websites, and public utterances. This applies to government, non-governmental organisations, and commercial, corporate, educational, and civil bodies.

A World Integrity Service would meet this need by certifying the integrity, after close independent examination, of the body and what it has to say, especially in regard to catastrophic risks. Such a certificate, with a formal Integrity Seal, would be a clear sign to the world public of the organisation's trustworthiness. Trustworthiness over time could be recognised by higher categories of seal – bronze, silver, and gold, for example – just as the cumulative performance of vendors on eBay is attested by their clients. Equally, any breach of trust or deliberate deception by the organisation would lead to the downgrading, suspension or even cancellation of its right to use the seal.

Any organisation bearing such a seal will be at a considerable advantage over others when it comes to attracting investment, staff, and clientele (or voters). Though such evaluation and certification may be expensive, it is a price which any honest body will be glad to meet. A clear distinction will be created in the public eye between bodies that earn the 'truth and integrity' seal – and those which cannot or dare not meet the test.

What You Can Do

- Appreciate that money, politics, religion, and entertainment are all belief systems and need to be reinforced with reliable facts from science if they are to be of greatest service to us.

- Be extremely careful how you spend your imaginary dollar. It has very real consequences.
- Buy from, invest in, and work only for companies that value the long-term survival of you and your grandchildren.
- Vote only for only political parties and leaders who value your survival – and that of your grandchildren.
- If you are a person of faith, do whatever you can to lead your co-believers to a position where the survival of humanity and our descendants, and the caring stewardship of the Earth, become paramount articles of faith and morality.
- Patronise entertainment that presents a more creative, co-operative, and less destructive narrative about humanity and how we solve our problems.
- Encourage your children to play games and watch shows that create, renew, and restore, rather than kill or destroy; understand that play is rehearsal for life and how children play may ultimately decide their fate in real life.
- Do not spread conspiracy theories or unvalidated claims you find on the internet, as this only widens the harm they do, may kill people, and reduces everyone's chances of survival. Unless they are evidenced by science, ignoring them is the best policy.[23]
- Support a World Truth Commission to expose misinformation by various groups and interests and a World Integrity Service to validate the trustworthiness of commercial and civil organisations.

12 WHO ARE WE, REALLY?

Without prompt and universal action, *Homo sapiens* is on track to earn the unhappy distinction of becoming the Earth's first species to use its intelligence to destroy itself.

Our intelligence has been great enough to encompass the Stone and Metal Ages, and the agricultural, industrial, and computer revolutions – but not, so far, to mitigate the risks which our exploding numbers and overuse of material resources have unleashed. As a species, we are technologically adept – but far from wise.

Indeed, the desirability of intelligence as a successful evolutionary trait is not at all clear. Jellyfish have been around for 500 million years or more,[1] and while not overly intellectual beings, they seem far better adapted to the demands of long-term persistence on Earth than humans. They will probably outlast us, as will fungi, microbes, algae, and other more 'primitive' yet persistent forms of life.

The time has come to question whether a species so bent on its own destruction, as well as that of other species, and on the fouling of its planetary cradle even merits the description 'intelligent', let alone 'wise'. As described in the opening chapter of this book, wisdom generally consists in the ability to look into the future on the basis of past experience and close current observation, discern a potential threat, and then act to nullify it – or even benefit from it. The birth of our kind, around a campfire that shielded us from predators was, quite likely, the watershed event in our rise – and the single act that separated us from other primates. It marked the moment our brains began to swell and births became difficult owing to the expanding size of the skull, necessitating a more tightly bonded society with the ability to share big ideas.

By 1758, we humans were so pleased with our apparent mental superiority over nature that the father of natural history, the Swedish scholar Linnaeus, felt able to name our species *Homo sapiens*, or wise man.[2] Wisdom, he felt, was clearly the attribute which sets us apart. It is intriguing to reflect what he would do today, were he able to board a time machine and consult his fellow naturalists about the appropriate nomenclature of humankind today.

When we study how modern humans behave, we see that it is anything but wise. We have wiped out two-thirds of the world's large animals. We are losing water, topsoil, fish, and forests at appalling rates. Every day, we poison everyone and everything on the planet. We are constructing weapons able to obliterate us many times over. We are shaping a climate that can render the Earth largely uninhabitable within the next three or four generations. We are building clever devices over which society has no control, but which can decide our future. We throw half our food away and wreck the planet trying to grow more. We unleash new plagues every few years and spread them worldwide like wildfire. Are any of these the acts of a wise, or even an intelligent, species?

Humans are marvellously inventive. Our minds are an endless fount of creative ideas – artistic, scientific, technological, social, and practical. But does that equal wisdom? Birds and apes use tools too – but do not employ them to endanger their own future. We are so enthralled by our latest invention, be it a steam engine, an aeroplane, a power station, or a smartphone, that we seldom bother to ask ourselves what could possibly go wrong. And, of course, commerce and advertising constantly reassure us that nothing *can* possibly go wrong, because that would be bad for business. Indeed, denial of catastrophic risk appears to be as hardwired into our social make-up as the need to innovate.

One of the chief obstacles to wise, universal action to overcome these threats is our self-admiration and complacency, and the blind overconfidence they endow us with. Insisting that we are a wise creature in the face of mounting evidence that we are crazy enough to risk destroying ourselves is a like a dangerous drug fantasy. It leads to baseless optimism about our ability to invent our way out of a universal crisis. Something urgent must be done to bring humans to a more realistic understanding of the creatures we have become.

In a letter to *Nature* in 2011,[3] I first proposed that *Homo sapiens sapiens* should formally be renamed and reclassified scientifically – not just because our existing name is a scientific nonsense in the face of all the evidence, but more especially because of the clear signal this would send to humanity generally about the unwisdom of our present behaviour. I developed this idea further in *Surviving the 21st Century*[4] and elsewhere,[5] proposing that our new name should be the subject of a worldwide debate, which engages the human population at large – not just a few scientists. 'A name is an essential part of who we think we are', I explained. Losing our claim to call ourselves 'sapiens' is a special form of shaming intended to awaken us to the realities of our behaviour.

The aim of such a worldwide debate would be to encourage all of humanity to better understand and think about our true nature, its attributes, and how these must change if we are to survive. And that includes replacing a name that, in itself, has become a threat to our future.

Changing the scientific name of a species is possible, but not easy. Scientists often fight to keep old species names, out of tradition or sentimentality, but given the threat that '*Homo sapiens*' now poses to its own survival these are inconsequential arguments. Technically, under the rules of international taxonomy, the name of a species can be altered if some, or all, of the following conditions are met:

- the discovery of new scientific attributes;
- changes in the common understanding of the species;
- changes are found in its evolutionary descent;
- there is a need to correct an error in its original name;
- lack of a type specimen (known as a holotype) representative of the whole species.

Homo sapiens sapiens should be renamed on all five of these grounds. Science has uncovered a great many new attributes of the species since 1758, including our evolutionary descent, our propensity for violent universal conflict, environmental pollution and extinction of other species, and the absence of a human holotype. The error in the original choice of name is abundantly clear.

If we are to 'fix our broken planet', we must know ourselves more truthfully – and act to control those behaviours that will otherwise imperil both ourselves and an Earth we can inhabit. However, the loss of the name *Homo sapiens* need not be a permanent 'demotion'. It can be earned back, provided humans fulfil certain criteria that together define us as a wise species.

As explained in the preceding chapters, 10 mega-risks, arriving at the same time, constitute the present human existential crisis. A wise humanity will do all in its power to prevent or minimise all of them. So, a suitable test of a wise humanity might include the following:

1. Have we eliminated weapons of mass destruction? How close are we to doing so?
2. Are we reversing climate change by ending the use of fossil fuels and returning the forests?

3. Are we ending our poisoning of the planet, by eliminating waste and replacing toxic chemistry with green chemistry?
4. Are we taking real steps to end the Sixth Extinction by rewilding the planet?
5. Are we putting wise women in leadership positions globally and heeding their advice?
6. Are we taking universal steps to lower human numbers voluntarily?
7. Are we recycling all minerals, nutrients, and water in a circular global economy?
8. Is society taking oversight and ethical control of dangerous technologies like nanotechnology, AI, and quantum computing?

The results of these tests will show clearly how wise or unwise humans are – and whether we are committed to a safer, healthier, more sustainable future, or to one that is dark, dangerous, and potentially deadly for most if not all of us.

Like the weather, the results of these tests should be broadcast on the evening news and appear on every smartphone and computer, on the back of cereal packets, and in every conversation and policy discussion. They should be examined in schools and universities, in homes, and in workplaces. Everyone should be able to see them, every day – because they define our very future. This can be achieved using a Human Survival Index (see Chapter 14).

As described in Chapter 1, humans probably first developed wisdom – the ability to perceive threats and opportunities clearly and then take considered action – more than a million years ago, when we first used fire as a defence against predators. Wise foresight has saved our species many times since. Wisdom has carried us through all the different ages of humanity, alerting us to dangers and enabling us to avoid or neutralise them, usually with technology and co-operation.

But we must first clearly understand the risks we face before we can solve them. With shared wisdom and co-operation, humanity can survive and prosper together. Without them, we go down in darkness together.

What We Should Do

1. Establish a worldwide debate on the future of humanity, and the name we call ourselves – because that name will define who we are and whether or not we can survive as a species.
2. Run the debate globally across media and social media, involving people from all backgrounds, beliefs, cultures, professions, and affiliations. Listen respectfully to others' views and discuss the factual evidence about human behaviour and its consequences that is furnished by science.
3. This debate must involve all humanity and cannot be restricted to the scientific community or other elites. It must be 'owned' by everyone, because we have all contributed to our present plight. However, science must supply the hard, objective evidence so that we human beings can realistically understand our true nature and behaviour, and their costs.
4. Whether or not humanity succeeds in agreeing on a new name is less important than the process of discussing our nature and the name we deserve. The aim is widespread understanding of our condition, the dangers it has created for all, and the urgent need to act together to mitigate them.

What You Can Do

*   Support a worldwide movement to rename humans truthfully – as the creatures we are, instead of the fantasy creatures we imagine, or would like ourselves to be.

- Start by discussing with your family, friends, and colleagues the defining features of human nature – and which ones most truthfully describe us.
- Tell scientists that you care – and demand to be involved.
- Put forward new names for humanity in the media and social media to advance global discussion. Explain your reasons carefully, based on evidence.
- Stimulate discussion in schools, universities, social clubs, parliaments, corporations, and round the family meal table.
- At all times keep the discussion civil and respectful of others, even if opinions differ widely, remembering that only by working together can humans survive our existential crisis.
- Remember at all times that the name chosen for humans is less important than gaining a realistic understanding of our nature – and the risks it brings. Our choice of name must reveal our true identity – and define the path by which we can truly call ourselves 'wise'.

13 AN EARTH STANDARD CURRENCY

In an age of existential emergency, when the future of civilisation depends on how successfully we manage to overcome the 10 global threats which are now bearing down on us all, it is important for humanity to share a common currency for dealing with them. We need not only a common understanding of the issues, but also a common means of exchange which enables us to rebuild and regenerate our damaged world.

The reasoning is straightforward: money is a figment of the human imagination. It exists only in the human mind, not in the real world. Yet we use it to develop, produce, exploit, damage, and destroy things that are real – like water, forests, oceans, the atmosphere, and life itself.

Being imaginary, money is also infinite in supply, in that banks and central banks can, as a rule, create as much of it as they choose, mostly out of thin air. Most of the money in circulation today is digital (more than 92 per cent). It consists of electrons in the computer systems of banks and trading houses and has no material existence. However, the planet on which we live is physical and finite. Its oceans, atmosphere, land mass, forests, soils, fresh water, minerals, and living organisms are all limited in extent and in terms of what they can yield for the survival of humans and other life on Earth. Science has long understood this. *If you use an infinite commodity, money, to exploit a finite planet, you will run out of planet long before you run out of money.*

This is the primary flaw in the present global monetary system: it looks only to our present need for 'wealth' and makes no investment in the long-term need for survival of either humans or life more generally. In economic terminology, it treats life itself as 'an externality'. In other words, through our monetary system, humans currently prize 'wealth' above our own survival. This is a worry. It needs to be understood and addressed with the utmost urgency.

To do so necessitates the creation of a medium of exchange which is based upon things that exist in the real world – not on things conjured from nowhere by banks and central bankers, or from the speculations and peculations of money market traders and brokers. The world of the twenty-first century needs a *real currency* with a trustworthy, reliable value, which people can use in their daily lives and businesses and which reflects the real planet and its many life-giving assets – these underpin the value of this currency. It is a currency which encourages humans to change their behaviour in favour of sustainable, regenerative activity instead of exploitative, destructive activity.

This is the proposed Earth Standard Currency. Call it an Earthcoin, a Gaia or what you will. The case for an Earth Standard Currency is that 'money' is infinite in a finite world and this is leading to the wholesale destruction and/or overexploitation of the systems needed to maintain life on our planet – forests, fish stocks, arable soils, fresh water, the climate and atmosphere, the oceans, wildlife, planetary and human health, and so on.

The volume of currencies presently in circulation is determined by trading banks, central banks, and governments, not by anything real. It is set according to their judgement of 'what the market will bear'. Previously, when the volume of money or the price of stocks or commodities exceeded this credibility limit, hyperinflation or a 'crash' usually ensued, so they tend to be cautious – but every now and then, they release a flood of new currency,

magicked from nowhere. Past cases when excessive speculation drove the theoretical value of commodities or
stocks too high, resulting in a collapse of public confidence, include the tulip crash (1634), the South Sea
Bubble (1711), Austrian and Weimar Republic hyperinflation (1920s), the Wall Street Crash and Great Depression
(1929), the wool boom (1950s), the Asian crash (1997), the
dot-com boom (2000), and the global financial crisis (2008).

The latter, many people will remember, was caused by
money plucked by banks out of thin air, then lent to unreliable ('sub-prime') borrowers. When the failure of these
bad loans became inevitable, they were then bundled and
sold on as 'derivatives' to people who did not understand
what they were buying. When confidence in these derivatives finally collapsed, the world financial system fell into
chaos and several major banks failed. The situation was
eventually stabilised by central banks plucking a further
US$3.5 trillion out of thin air to pay for the bad debt created
by the banks' bad loans. Thus, imaginary money was used
by central banks to pay off the imaginary debt created out
of nothing by banks, lenders, and speculators in the first
place. This pecuniary prestidigitation was dressed up in
the fancy term 'quantitative easing'. In other words,
a speculative oversupply of 'money' had led to public loss
of confidence in its value.

It follows from this experience that it is desirable for
money to have a value that is based in the real world,
which is not the plaything of gamblers and speculators,
and which cannot be created at whim by irresponsible
lenders or internet sharks, as is the case with so-called
'cryptocurrencies'. An Earth Standard Currency, the
value of which is based on actual assets, physical and
ecological, for sustaining life on Earth meets such
a need.

There is no precedent for a currency whose value is
based on the real value of the planet. However, there are

useful examples. Gold, for example, has been used to underpin rates of exchange for over 2,700 years. Prior to the 1930s, governments based the value of their currencies – such as sterling and the US dollar – on the 'gold standard'. This was based on a fixed value for gold, set by government, and involved the holding of large, unwieldy gold stocks like those kept in the vaults of the Bank of England or Fort Knox. This enabled countries to settle their debts by large physical gold transfers between them – a cumbersome and risky process.

In the 1930s, gold was replaced with 'fiat money' in which the value of the currency was allowed to float at will against other currencies, underpinned by a guarantee from governments. This system prevails today. Its chief shortcoming is that it allows the creation of 'imaginary money' and feeds speculation by money marketeers, giving most of the world's currencies the same value as a roulette chip. Bitcoin and its imitators are examples of totally imaginary currencies, manufactured out of thin air or even less, and then traded among people who are gambling they can sell it on for a higher price before enough people recognise it for the Ponzi scheme it is – and the whole house of cards collapses. The dangers of this system failing have greatly increased with the widening use of AI in money market trading, raising the risk of 'computer error' precipitating a collapse.

For nearly three millennia 'money' has been so useful a way of exchanging various goods, ideas, and services that we now cannot do without it, at least for the time being. However, money works best when its value is known, reliable, and trustworthy – not fluctuating every second.

So, the alternative to our present untrustworthy and destructive systems is to create a universal currency whose value is tied to the combined value of the assets of the planet which are used to support life – its air, water, soil, forests, etc.: the Earth Standard Currency. While these assets change over time – mostly as we degrade or improve them – they do not do so with the volatility of present currencies. Moreover, their changes can be measured scientifically – the composition of the atmosphere, the area of forest, the rate of soil loss, the availability of fresh water. If it is made up of these factors, the value of an Earth Standard Currency will be far more stable and secure, though it will change with time.

A second precedent is the European euro, the official currency in 19 of Europe's 27 member states and the world's second largest reserve currency after the US dollar. Introduced in 1992, the euro is now consistently worth more than the US dollar and its volume in circulation exceeds it. The value of the euro is set by the European Central Bank based on a fixed conversion rate for each member country's national currency. However, its flaw is that these component currencies are subject to the same volatility as the US dollar and so are vulnerable to overproduction, speculation, or national economic mismanagement. The success of the euro demonstrates that it is possible to create a new currency which serves more than a single country and, in a globalised world, it represents a broad model for introducing an Earth Standard Currency.

Undoubtedly, an Earth Standard Currency would succeed for the same reasons as the euro, by offering greater stability and security for users, savers, investors, and traders in a world where the values of various other currencies are set by lenders and gamblers as well as by the financial management, good or bad, of participating countries. Since most governments in the world are either self-appointed or else the product of corrupt political processes, it follows that the present value of the world's currencies lies chiefly in the hands of the power-hungry and greedy and has little relationship to the needs of either people or the planet.

How Would It Work?

The Earth Standard Currency is, in essence, an index compiled by scientific measurement of all the main ecological systems and physical resources that sustain life on Earth, including human life. This index would involve an algorithm compounded from the objective measurements of key planetary assets such as the volume of fresh water available, the state of the atmosphere required for a stable climate, the extent of forest cover, the rate of global soil loss, the extent of pollution and poisoning of the biosphere, the number and condition of living species, and so on. These measurements constitute a realistic appraisal of the ability of the Earth to sustain life – and whether that ability is improving or deteriorating under human 'management'.

In other words, an Earth Standard Currency is an 'index of indices', balanced against one another to reflect the realistic state of the Earth's resources and life support systems. Achieving such a balance is no trivial undertaking scientifically, but there are numerous examples of concepts which can aid in its formulation. These include:

- the Global Footprint Network[1];
- the Global Boundaries concept propounded by Rockström, Steffen, et al.[2];
- the Global Prosperity Indices developed by numerous scholars as a better way than GDP to measure national well-being[3];
- the Gross National Happiness Index of Bhutan, which comprises sub-indices of living standards, education, health, environment, community vitality, time-use, psychological well-being, and good governance[4];
- the Threatened Species Index developed in Australia[5];
- the Global Fishing Index.[6]

The key points in favour of an Earth Standard Currency are:

1. It is based on real, tangible Earth systems' assets, constantly monitored and recorded by science. It is thus not subject to speculation, selfish manipulation, or mismanagement by governments, banks, or money traders.
2. Its value will fall over time as Earth systems deteriorate, and will rise again as they improve or are regenerated by human action. This means the value of your home or your pay packet will rise and fall along with the value of the Earth Standard Currency, providing a modest but real incentive for all users to work harder to improve the habitability of planet Earth.
3. An Earth Standard Currency will foster the development of the circular economy and the 'ideas economy' (based on non-material goods and services), thus providing for economic growth without ecological penalty.
4. People and investors will trust it, and adopt it readily, because it is independent from the machinations and greed of national governments and private speculators.
5. It provides a global currency for a global world, which sends a clear signal about the 'liveability' of that world for all. It represents a universal human aspiration to survive and prosper within the natural bounds set by a habitable planet.

An Earth Standard Currency will need a central authority to manage and administer it and oversee the scientific integrity of its indices. It is recommended that this body is created as a special institution under the UN – say, by combining the World Bank and International Monetary Fund – and that its creation should be an important feature of the proposed Earth System Treaty.

Possible Objections

No doubt there will be vocal objections to the creation of an Earth Standard Currency, chiefly from economists and from those sectors that imagine they have the most to lose from the stabilising of world currency around a scientific rather than a market metric that they cannot play with. Some individual countries may feel threatened by an Earthwide currency because it reduces their ability to manipulate the world economy and the lives of others for their own selfish benefit. Inevitably, people who live by gambling on the value of currencies, including the world's banks, will moan and groan about it. But their ability to survive on the Earth is at risk too, though they are too short-sighted to see it.

However, the euro is evidence that a stable, universal currency with a known and predictable value will appeal strongly to most people. It will give their savings a safer haven, reduce the cost of money market antics, and will also make trade and exchange rate risk far less of a gamble for international buyers and sellers of goods, ideas, and services – if not eliminate it altogether.

English naturalist and broadcaster Sir David Attenborough is famed for his statement that 'Anyone who thinks that you can have infinite growth in a finite environment is either a madman or an economist.'[7] However, economists have yet to absorb this basic truth. An Earth Standard Currency offers them a chance to participate in and lead in the creation of a real currency that will restore the planet's resources for life instead of consuming, wasting, and destroying them. Economists, working with scientists and statisticians, can assist in the design of workable indices, and the correct balance between them, to create a trusted world currency for everyone that can help repair our broken planet.

Furthermore, an Earth Standard Currency can both help to stabilise human use of the Earth's resources and at the same time contribute to economic growth by encouraging the development of the 'ideas economy' in which wealth is based on non-material goods and services such as software, the arts, sport, entertainment, science, technology, communication, human services, etc. The 'ideas economy' provides jobs, growth, and profit but does not squander the Earth's finite resources and pollute, as an economy based on steel, cement, oil, coal, agriculture, forestry, fishing, lost water, etc. does.

Market traders need not fear an Earth Standard Currency as they can continue to trade in the many other national currencies or stocks that will remain. However, they will have no power to influence its value for selfish purposes.

There is at present a fierce debate in academia and in the public domain about whether 'capitalism' is to blame for the destruction of the planet and if it should be discontinued in favour of some other socio-economic system. It is doubtful that either side in this debate will ever win the argument, as the last 200 years have settled nothing politically. It is certain that any attempt to undo the present monetary system would be more fiercely opposed than even efforts to prevent fossil fuels from wrecking the climate, as so many people have a vested interest in it and they dominate world economics and politics. The creation of an Earth Standard Currency sidesteps this argument completely, offering a pathway to a stable global medium of exchange which encourages the regeneration of the Earth along with continued, but far less destructive, economic growth.

Furthermore, being independent of the selfish influence of national governments, transactions denominated in an Earth Standard Currency can be levied by a small amount (e.g. 0.1 per cent) to help fund projects of global urgency and significance such as the Sustainable Development Goals (SDGs), global reforestation, revegetation and rewilding (for instance, the UN-REDD programme), ocean cleanup, carbon farming, and water recycling. This will create an international fund for projects to repair and regenerate the planet.

What We Must Do

The UN, academia, national governments, and central banks should establish an international authority to introduce and oversee a universal Earth Standard Currency on the principles outlined above. This could be created by merging the existing World Bank and International Monetary Fund and giving the merged body a new charter for a world currency. An Earth Standard Currency could become a central part of the proposed Earth System Treaty.

The value of the currency should be based on a formula that embodies the scientific measurement of all the Earth's main life-sustaining assets – the soil, the oceans, the atmosphere, freshwater resources, forests, the condition of wild species, and so on. This value will change only as these assets degrade or improve under human management, leaving no room for currency speculation.

What You Can Do

- Ensure you understand how the present money system damages and destroys our world and reduces our long-term prospects of survival in it.
- Ensure you understand how your own spending helps to destroy or heal our planet, and defines the future of humans on it. Spend more wisely.
- Be an advocate for an Earth Standard Currency, to keep your savings secure, your pay stable, and to invest in your children's future on a safe, habitable planet.

14 TOOLS FOR REPAIRING THE EARTH

In the course of this short book many actions have been described which are essential to creating a safe, habitable world for humans to occupy in the future, alongside the rest of nature. Some of these actions – like eliminating carbon emissions or nuclear weapons – have begun, or are just beginning, to be implemented. Others have yet to start. However, there is as yet no overarching, integrated programme of action to save humanity and a habitable Earth.

Above all, human civilisation needs a coherent plan for its own survival that is acknowledged by all nations and corporations, and to which they willingly and sincerely contribute. It needs to be practical, achievable, fair, and likely to receive the support of most citizens. And it should make none of the catastrophic threats worse and be integrated across all the threats. Such a plan will not arise on its own. It needs universally agreed instruments, or tools, to set it in motion. The main ones are summarised below.

An Earth System Treaty

An Earth System Treaty is a global accord, negotiated, signed, and ratified by all the countries of the Earth, under the auspices of the UN. Once declared, it should also be available for voluntary signature by individual citizens, corporations, non-government bodies and other groups, organisations, and agencies worldwide, so that all may have a chance to affirm their commitment to the whole planet and the safer future the treaty provides for humans and other forms of life.

The purpose of the treaty is to provide an international legal framework for protecting and restoring the Earth system so that humans can inhabit it indefinitely. It is a global instrument for achieving human survival and well-being and for caring for the planet that affords those benefits. It sets standards, objectives, and boundaries that all can follow and abide by.

Several forms of treaty have been proposed,[1] but none has so far managed to encompass all 10 of the catastrophic threats that constitute the human existential emergency or to integrate their solutions. Most versions refer chiefly to climate change and environmental decline, but overlook questions such as population, food, nuclear weapons, ultra-technology, and misinformation. Unless all 10 mega-threats are included, human survival will remain in doubt.

A truly effective Earth System Treaty will embody:

- a universal ban on nuclear weapons;
- an international plan to combat climate change;
- an international plan to restore forests, soils, fresh waters, oceans, the atmosphere, and biodiversity to stable, sustainable levels;
- an international agreement to operate a circular economy and end waste;
- a plan for a renewable world food supply sufficient for all;
- a plan to end chemical pollution in all forms;
- a plan to voluntarily reduce the human population to a sustainable level;
- a global plan to anticipate and prevent pandemic disease;

- a Global Technology Convention to oversee the safe development and introduction of advanced sciences and technologies and minimise harms;
- a World Truth Commission;
- an Earth Standard Currency;
- all 17 of the Sustainable Development Goals.

The concept behind the treaty is that all signatory nations bind themselves to accepting its standards and promoting them internally, and to others. Inviting individuals to sign the treaty will call on the enormous power and genius of grassroots humanity to implement it – as well as signalling that individuals, as well as nations and corporations, are also expected to play their part in saving humanity.

Above all, an Earth System Treaty will seek to keep human activity within safe boundaries set by the operation of the Earth system itself, as shown in Figure 14.1 produced by the Stockholm Resilience Centre. Humans are already in the danger zone in respect of destruction of the biosphere, climate change, land clearing, and excessive chemical flows such as nitrogen and phosphorus.

A Renewable World Economy

A Renewable World Economy is one where money is not used to destroy, but rather to rebuild, restore, and renew. The principles have been explained earlier but broadly it consists of:

- a circular or doughnut economy implemented on a global scale, recycling as close to 100 per cent as possible everything that humans use;
- an economy whose growth relies not on physical materials but on ideas;
- an economy that embraces the ideas of equity, social justice, and fairness as well as sustainability;
- an Earth Standard Currency whose value reflects the ability of the Earth to sustain life (see Chapter 13).

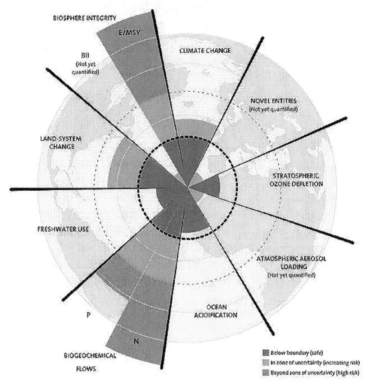

Figure 14.1 The concept of boundaries or limits to human activity is essential to our survival. This diagram shows how humans are placed in terms of our impact on the key factors that control the Earth system. In five of these we have already entered the danger zone. Such boundaries need to be part of an Earth System Treaty. Credit: J. Lokrantz/ Azote based on Steffen W, Richardson K, Rockström J, et al. Planetary boundaries: Guiding human development on a changing planet. *Science*, 2015, 347 (6223). http://doi.org/10.1126/science.1259855.

Biosphere integrity (BII) – the state of the environment; E/MSY indicates the rate of extinction; P – phosphorus; N– nitrogen.

Many other economic reforms too numerous to describe here are also needed, in particular the ascribing of monetary value and legal rights to the natural world and its processes. Many of these are articulated in the United

Nations Sustainable Development Agenda for 2030,[2] and the 17 Sustainable Development Goals and their 169 targets.[3] All are essential pillars of a *sustainable* world economy.

New Human Rights

Although it is occasionally updated, the 1948 Universal Declaration of Human Rights[4] and its 30 articles is essentially a document based in the mid-twentieth century, when it was first conceived. It makes insufficient allowance for the spectacular explosion in the impact of advanced technologies, overpopulation, overconsumption, global pollution, the destruction of the natural world, and other developments that affect the freedoms and rights of all individuals. It is therefore necessary to augment the declaration in several ways, so that it is a more useful device for securing human survival (and thus, the right to life) in the present age, by including the following:

- a *right not to be poisoned*, including by nanoparticles and novel, untested chemicals;
- a *right prohibiting the mass surveillance of populations*;
- a *right prohibiting the creation of dangerous biological entities*, which may cause future pandemics and unknown diseases or bioweapons;
- a *right prohibiting the use of killer robots* against civilian populations.

As previously noted, the mere existence of a right does not prevent its being broken or infringed – but it does establish a universal standard of behaviour, help create legal avenues for redress, publicise the issue, and encourage all to adopt it; it also leads to public exposure and shaming of those who breach it.

An important consideration in reviewing human rights is whether the Earth system itself should be accorded legal rights to protect it from abuse by unscrupulous humans. The Maori of New Zealand have successfully argued in law that the Whanganui River is a living entity, entitled to the same legal protections as a person.[5]

A Global Technology Convention

In recognition of the fact that human numbers and demands are now so vast that almost any technology created to satisfy them will have a negative impact on the Earth and its ability to sustain life, we need a Global Technology Convention requiring that *all* technologies – especially those that are comparatively recent and unknown – have both their benefits and their harms assessed independently, objectively, and publicly. Past experience shows that people who develop, make, sell, or use certain technologies are inclined to talk up the benefits and play down the harms, and so cannot be trusted to tell the whole truth about them. A mechanism for societal oversight is now essential – such is the scale of the risk to all humanity that powerful new technologies now present. A global agreement and an independent agency for reviewing novel technology, akin to the Intergovernmental Panel on Climate Change (IPCC), is the only way we can avoid repeating the catastrophic mistakes of the past or creating fresh disasters in future.

A World Truth Commission

Just as the Truth and Reconciliation Commission[6] took vital strides in healing and bringing together the new nation of South Africa, the world urgently needs to face the honest truth about the catastrophic threats it is confronted with in order to organise global action against them. A World Truth Commission, empowered to investigate dubious claims by powerful players – corporate, social, or political – report its findings, and expose the deceptive or mendacious, is an essential step towards building the global trust critical to human survival.

A World Truth Commission could take the form of a tribunal similar to the International Criminal Court (ICC), with powers to investigate any statement or claim made by an influential public figure or organisation, in particular, those that may have an influence on catastrophic and existential risks and the public perception of them. It would be staffed by legal, ethical, and scientific experts and draw on academic expertise globally. Its findings would be reported publicly. Individual citizens, their organisations, government agencies, and countries would all have the right to refer cases of misinformation and deception related to catastrophic risks to the commission for fact-checking. Organisations or individuals found to use misinformation in ways likely to increase the risk of death, harm, and destruction should be referred for prosecution to the ICC, in the same way as war criminals are today.

This concept of a World Truth Commission is coupled with that of a World Integrity Service providing independent validation of the integrity and truthfulness, after close independent examination, of any applying organisation and what it has to say, especially as it bears on catastrophic risks. A World Integrity Service Seal, for example, could help inform internet users whether particular websites can be trusted, consumers whether corporations and their products are to be relied on, and voters whether governments or political parties can be trusted to tell the truth.

A Human Survival Index

In order to raise global awareness, the Council for the Human Future has proposed the creation of a Human Survival Index displaying the scale of threat to humanity from the combined existential risks we now face, and which are of our own making. This is an easily understood tool to inform the public of the need for common, resolute, planet-wide action to mitigate these risks.

The Human Survival Index is a scientifically measured report card on the human future. Consisting of 10 component indices, each describing one of the major risks, it informs civil society and government whether the overall threat to civilisation is improving or growing worse – and whether our remedial actions are working or not.

Like other familiar indicators such as the stock market index, the money markets, the Doomsday Clock, the Global Boundaries diagram, the Global Footprint, and the daily weather forecast, the Human Survival Index would be publicised worldwide on news broadcasts and across all media, including social media, with the aim of informing and reinforcing global action by individuals as well as by governments and organisations.

Stewards of the Earth

A Stewards of the Earth Programme is a global initiative to repair ruined landscapes, replant forests and grasslands, restore natural waterways and wetlands, lock up carbon, and regenerate nature. Its workforce consists primarily of willing farmers and indigenous people, who have a deep understanding of their own local environments and are funded through a global levy on (a) military budgets and (b) food and natural goods.

With corporate agribusiness throwing hundreds of millions of small farmers off their land and replacing them with industrial food systems, the need to retain their skills, knowledge, and love of country for restoration of the planet is urgent.

A Stewards of the Earth Programme will immediately create a locally skilled workforce for environmental repair and restoration, prevention of extinction, and prevention of new pandemic diseases, as global food production moves progressively into cities and the deep oceans (see Chapter 8).

The aim of the programme is to restore and rewild the global natural environment so that it is able to sustain human civilisation, together with the Earth's existing biodiversity, into the future.

A Year of Food

Designed to teach the young how to eat healthily and sustainably, a Year of Food would be embedded in every class in every junior school on the planet and *every* subject, including sport, is taught through the lens of food. Its unifying theme is how to choose or grow healthy, sustainable foods and the aim is to build global demand for a renewable food system to replace the failing industrial system. Children, in turn, will help educate their parents about healthy, sustainable ways to eat. The Year of Food could be managed and promoted by agencies such as the UN Food and Agriculture Organization, the World Food Program, and UNICEF with the support of non-governmental organisations such as Oxfam and CARE.

A Global Population Plan

Since population is the mainspring of all the major threats to the human future, it is of overwhelming importance that humans control our numbers before nature does it for us. A World Population Plan, embracing voluntary family planning, education, materials, and advice, is indispensable to a safe, sustainable human future. It must be backed with adequate investment to ensure free family planning is available to every woman and her family on Earth.

The plan should map out targets and pathways to achieve the ideal sustainable human population at the earliest time possible, in order to avoid catastrophe caused by people outrunning their resources, migrating en masse, and fighting over them. Such a disaster is clearly foreseeable, and planning for its avoidance is now both essential and urgent.

Conclusion

Many other tools for building a safe future for humanity have been proposed, but these seem among the best and most urgent. They will have their critics – but the challenge for any critic is to propose a better solution, not merely to criticise. Above all, any plan for the human future must address all the threats, make none of them worse, and provide practical, integrated solutions for the complex problem of human survival.

15 THINK LIKE
A HUMAN, ACT LIKE
A SPECIES

Humans are at the threshold of the most significant moment of our entire existence: learning to think as a species. This is the linking of human minds, values, information, and solutions at light speed and in real time around the planet, via the internet and social media.

Thanks to the spread of the internet and social media, people are able, for the first time, to communicate across the barriers of language, race, nationality, religion, culture, gender, and socio-economic status that have long divided us.

Global thought may be arriving in the nick of time, opening the way to solving some of humanity's greatest threats – including climate change, famine, global poisoning, weapons of mass destruction, environmental collapse, resource scarcity, pandemic disease, dangerous new technologies, and overpopulation.

While the internet contains much rubbish and malignancy, it also carries huge amounts of goodwill, generosity, trustworthy science-based advice, and practical solutions to problems – as well as people joining hands in good causes. Users just have to be able to choose wisely between the two.

There is a vivid parallel that applies to everybody. In the second trimester of a baby's gestation a marvellous thing happens: the nerve cells in the embryonic brain begin to connect – and a mind is born. An inanimate mass of cells becomes a sentient being, capable of thought, imagination, memory, logic, feelings, and dreams. Today individual humans are connecting, at light speed, around a planet – just like the neurons in the foetal brain. We are in the process of forming a universal, Earth-sized 'mind'.

A higher understanding, and potentially a higher intellect, is in genesis – one that is capable of thought, reason, and resolute action to counter the existential emergency that is building up around us. Humans are learning to think at *supra-human* level by applying millions of minds simultaneously to issues in real time, by sharing knowledge freely, and by generating a faster global consensus on what needs to be done to secure our future.

Today, irrefutable scientific evidence confirms that humanity faces 10 catastrophic risks, the result of our burgeoning population and overgrowth in its demands on the Earth's natural resources and systems. However, practical solutions to all of these problems exist – and are capable of being shared universally and instantly.

The problem we face is that some governments, big corporations, and industrial/lobby groups are loath to act. Blindly, they place short-term self-interest above the interests of the human species in sustaining our existence on the planet. Some people may believe that, by serving their own interests, they can somehow survive when the crash comes; they cannot. Others may seek short-term advantage in spreading lies, fantasy, and disinformation about the reality of our plight. They too will not avoid their fate.

The internet is showing that their time is up. The energy debate, and the way it is accelerating the transition to clean energy, is a perfect example of this. The growing worldwide resistance to toxic food and consumer products, from the United States and China to Europe, Australasia, and South America is another case in point. The global response to the Covid-19 pandemic is a third case, and the desire to restore the environment a fourth. All prove that humans are capable of acting wisely together, at world-wide scale.

What will take over from failing national governments and transnational oligarchies will not be a 'world government', as some people imagine. It will be a human species that shares thoughts, ideas, values, and solutions at light speed – an Earth-sized democracy, capable of disciplining any government, body, or corporation that tries to put self-interest above the supreme human interest in surviving.

In 2021, over 5.1 billion people, of a world population of 8 billion, used the internet.[1] It is expected that by 2030 everyone will be online. For the first time in history a conversation among the whole of humanity will become possible – and what more urgent and relevant topic than the survival of human civilisation and of our species?

Through the internet, young people and older people alike are reaching out to one another in real time, across the divides of race, nationality, ethnicity, language, religion, generation, gender, politics, socio-economic status, and prejudice. They are learning how similar we all are. How many things we share. How we can 'like', help, support, and depend on one another.

They are also learning how deadly are the prejudices, the ignorance, the fears, and the hatreds of our predecessors towards other humans – the things which bred the wars of past centuries. And how utterly pointless.

The antidotes to ignorance and fear are knowledge and understanding. The internet is capable of supplying both. People just need to be able to discriminate between what is good for humanity – and what is not – and between what is true and trustworthy and 'fake news'. We need to become 'informed consumers' on the internet, as we are in choosing foods or any other product for safety, health, and sustainability. Above all, we need to hear more women's voices about the human future and our needs as a civilisation.

The antidote to despair is action. In this book there are actions aplenty, great and small, for governments, corporations, and individual citizens. There are more to be found, by those who seek diligently in the universal human mind that is forming online and who are also willing to learn from nature, which has been practising survival for far, far longer than we have.

The antidote to blindly walking into catastrophe is wisdom. We have practised it for over a million years, identifying the threats to our future – and using our intelligence, inventiveness, and co-operation to counter them. We must now practise wisdom in earnest, together, to escape the existential emergency our lack of foresight has bred.

What We Must Do

We must consider and adopt a universal plan for human survival that involves a worldwide effort to fix our broken planet. This can be incorporated in an Earth System Treaty, a legal framework agreed to and signed by all nations and implemented universally at global, national, and local levels (see Chapter 14).

In the words of former Australian Science Minister Barry Jones, channelling Abraham Lincoln, 'We must consecrate ourselves to the unfinished work of saving Planet Earth, our home, where our species ... lives and depends for its survival.'[2]

What You Can Do

- Be a leader at this greatest moment in human history.
- Resolve to play an active part in the most important development since our species evolved on Earth.
- Use social media and the internet to join hands with like-minded people all around the planet, to

overcome the divisions which nationality, race, religion, politics, gender, education, money, and class have created.

- Meet (personally or online) with people from other lands, beyond your immediate circle of friends, colleagues, and neighbours. Find out what issues and values you share, what challenges you face, what opportunities there are to help each other and work together on our common cause, and what solutions you can find together.
- Be a willing volunteer. Set up or work with a group dedicated, in some way, to human survival – whether by repairing the natural world, improving health and knowledge, sharing, showing generosity, helping, feeding, and supporting.
- Do something every day to support the 17 Sustainable Development Goals.[3]
- Use your voice on the internet and in society to speak out against weapons of mass destruction, fossil fuels, toxic chemicals, land clearing, cruelty, discrimination, injustice, and all the other threats discussed in this book.

In Conclusion

Of course, the ball of rock, liquids, and gases that we know as the Earth is not 'broken', as the title of this book might imply. It is the human-habitable Earth which is broken – and which we must fix, for all our sakes. As the father of Gaia theory, the late James Lovelock, cautioned, 'we are now so abusing the Earth that it may rise and move back to the hot state it was in 55 million years ago and most of us and our descendants will die'.[4]

The Earth will continue, with or without humans. It does not care. Life will continue, with or without humans. It does not care. Life will survive the heat; the poisons; the extermination of individual species, genera, and even entire families of creatures; and the universal vitiation of air, water, and land. We humans will become simply another failed evolutionary experiment that lingered for a while – and then, thoughtlessly, engineered our own departure because we lacked the wisdom to avoid it. Should a scientific intelligence ever arise again on Earth, it may discover our fossilised bones amid the wreckage left by the fossil fuels and poisons we ourselves unearthed. The Earth has been unsuited for anything other than microbes in the past – and it may become so again if we continue to mismanage it.[5]

Unless we act. The actions enumerated in this short book are a start. They do not amount to all the answers, certainly – but many of them. Hopefully, they will inspire you and others to develop more and better solutions. They may not prevent a crisis, even a collapse, driven by our excessive numbers and demands. But they can limit the damage that results, the scale of death and suffering, if most of us can act together, promptly, and with goodwill.

Provided our solutions do not, in turn, unleash fresh threats, they can save many millions of lives and spare human suffering on a universal scale. And if we can bury all the foolish differences – tribal, political, religious, social, economic, cultural, national, generational – that now divide us, we can join hands and go forward as one people on a single planet, with a common goal.

> *In action lies hope, and in hope lies the future for our kind. Every other path leads to darkness.*
> *So let us act. Now. Together. Wisely.*

NOTES

1 Existential Emergency

1 Brain CK. Were our early ancestors murderers and head-hunters? A prehistoric detective story. *Quest* (Academy of Science of South Africa),2009, 5 (2), 19. https://journals.co.za/doi/epdf/10.10520/EJC89696.

2 Brain CK & Sillent A. Evidence from the Swartkrans cave for the earliest use of fire. *Nature*, 1988, 336, 464. https://doi.org/10.1038/336464a0.

3 Cribb JHJ. *Surviving the 21st Century: Humanity's Ten Great Challenges and How We Can Overcome Them*. Springer, 2017.

4 In this book, a catastrophic risk is one that presents an overwhelming threat to much of humanity and civilisation; an existential risk is one with the potential to eliminate the human species.

5 Council for the Human Future. *Delivering the Human Future*, March 2021. https://humanfuture.org/human-future-report.

6 Common Home of Humanity. www.commonhomeofhumanity.org/ (accessed 23 August 2022).

7 Magalhaes P, Steffen W, Bosselmann K, Aragão A, & Soromenho-Marques V. *A Safe Operating Space Treaty*. Cambridge Scholars, 2016.

8 Jones BO. *What Is to Be Done?* Scribe Press, 2021, p. 352.

9 See Cribb JHJ. *Food or War*. Cambridge University Press, 2019.

10 See Cribb JHJ. *Earth Detox*. Cambridge University Press, 2021.

2   Extinction ... or Survival?

1 CBC Radio. This 'really strange' spiralling sea creature may be the longest animal in the ocean, 16 April 2020. https://bit.ly/3QAiy7A.

2 Estimates put the world chicken population at 33 billion in 2020. See Shahbandeh M, Global number of chickens 1990–2020. *Statista*, 21 January 2022. www.statista.com/statistics/263962/number-of-chickens-worldwide-since-1990/.

3 Ceballos G, Ehrlich PR, Barnosky AD, et al. Accelerated modern human–induced species losses: Entering the sixth mass extinction. *Science Advances*, 2015, 1 (5). http://doi.org/10.1126/sciadv.1400253.

4 WWF. Living Planet Report 2022. https://livingplanet.panda.org/en-us/.

5 Marchant N. 9 of the most shocking facts about global extinction – and how to stop it. World Economic Forum, 2 November 2020. https://bit.ly/3QiTJO2.

6 Fleming S. What is ecocide and can it be prosecuted by the International Criminal Court? World Economic Forum, 1 July 2021. https://bit.ly/3Ac0SKh.

7 WWF, *Living Planet Report 2020*, p. 4.

8 Ceballos G. Extinction and the future of humanity. Council for the Human Future, March 2021. https://youtu.be/meizWp7jtUY.

9 Woolhouse M and Gaunt E. Ecological origins of novel human pathogens. *Critical Reviews in Microbiology*, 2007, 33 (4), 231–42. http://doi.org/10.1080/10408410701647560.

10 First draft of the post-2020 global biodiversity framework, Convention on Biodiversity, UNEP, 5 July 2021. www.cbd.int/doc/c/abb5/591f/2e46096d3f0330b08ce87a45/wg2020-03-03-en.pdf.

11 Plackett B. How many early human species existed on earth? *LiveScience*, 20 January 2021. www.livescience.com/how-many-human-species.html.

12   Cribb, *Surviving the 21st Century*.

13   UN Climate Change Conference 2021. Glasgow leaders' declaration on forests and land use, 2 November 2021. https://bit .ly/3pFDzma.

14   UN-REDD. www.un-redd.org/ (accessed 12 September 2022).

15   Wilson EO. *Half Earth – Our Planet's Fight for Life*. W. W. Norton & Co, 2017.

16   Cribb, *Food or War*.

17   Savory CA. Facilitating the regeneration of grassland. https:// savory.global/.

18   FAO. *The State of World Fisheries and Aquaculture*, 2020. www.fao.org /documents/card/en/c/ca9229en.

19   Gonzalez R. Decommissioned oil and gas platforms eyed for aquaculture use. *Aquaculture North America*, 3 August 2019. https:// bit.ly/3Kg8nTH.

20   WWF, *Living Planet Report 2020*.

21   UN. Climate action – initiatives for action, 2022. www.un.org/en/ climatechange/climate-action-coalitions.

22   Selgelid MJ. Gain-of-function research: Ethical analysis. *Science and Engineering Ethics*, 2016, 22 (4), 923–64. https://doi.org/10.1007 /s11948-016-9810-1.

23   Sustain. *The Sustain Guide to Good Food: How to Help Make Our Food and Farming System Fit for the Future*, 2013. www.sustainweb.org /publications/the_sustain_guide_to_good_food/.

24   For example, WWF, *Living Planet Report 2020*.

25   WWF. www.worldwildlife.org/ (accessed 12 September 2022).

26   Half Earth project. www.half-earthproject.org/ (accessed 12 September 2022).

27   Mission Blue. https://mission-blue.org/ (accessed 12 September 2022).

28   Jane Goodall Institute. www.janegoodall.org/ (accessed 12 September 2022).

29   Greenpeace International. www.greenpeace.org/international/ (accessed 12 September 2022).

30 Friends of the Earth International. www.foei.org/ (accessed 12 September 2022).

31 The Sierra Club (USA). www.sierraclub.org/ (accessed 12 September 2022).

32 Ethical Consumer. What is greenwashing?, 19 February 2020. www.ethicalconsumer.org/transport-travel/what-greenwashing.

33 Environmental Working Group. EWG's 2021 shopper's guide to pesticides in produce. www.ewg.org/foodnews/dirty-dozen.php; EWG's 2022 shopper's guide to pesticides in produce.www .ewg.org/foodnews/clean-fifteen.php.

34 Hunnes D, *Recipe for Survival*. Cambridge University Press, 2022.

35 WWF. WWF Armenia green living tips. https://wwf.panda.org/ wwf_offices/armenia/help_us/eco_help_living/ (accessed 12 September 2022).

36 Selinske M, Garrard, GE, Gregg, GA, et al. Identifying and prioritizing human behaviors that benefit biodiversity. *Conservation Science and Practice*, 2020, 2 (9), E249. https://doi.org/ 10.1111/csp2.249.

37 Monbiot G. @GeorgeMonbiot (8 April 2022). Almost every day I'm asked 'but what can I do?'. Twitter Page. https://bit.ly/ 3QLtCiX.

3   Resources for Living

1 CGRi. Five years of the Circularity Gap Report, 2022. www .circularity-gap.world/2022.

2 Worldometer. Global CO_2 emissions, 2022. www.worldometers.info /co2-emissions/.

3 Global Footprint Network. Earth overshoot day, 2022. www .footprintnetwork.org/our-work/earth-overshoot-day/.

4 CGRi. *The Circularity Gap Report*, 2022. https://bit.ly/3Ajvfx6.

5 Mekkonen MM & Hoekstra AY. Four billion people facing severe water scarcity. *Science Advances*, 2016, 2 (2), e150032. https://doi.org /10.1126/sciadv.1500323.

6 FAO. *The State of the World's Forests*, 2020. www.fao.org/state-of-forests/en.

7 UNCCD. *Global Land Outlook*, second edition, 2022. www.unccd.int/sites/default/files/2022-04/UNCCD_GLO2_low-res_2.pdf.

8 Wilkinson BH & McElroy BJ. The impact of humans on continental erosion and sedimentation. *Geological Society of America Bulletin*, 2007, 119 (1–2), 140–56 and Borrelli P, Robinson DA, Panagos P, et al. Land use and climate change impacts on global soil erosion by water (2015–2070). *PNAS*, 2020 117 (36), 21994–22001. https://doi.org/10.1073/pnas.2001403117.

9 Cameron D, Osborne C, Horton P, & Sinclair M. A sustainable model for intensive agriculture. Grantham Centre briefing note, University of Sheffield, December 2015. https://bit.ly/3QcEp4F.

10 While strong measures to prevent wind erosion have been taken in many developed countries, there is generally far less control over water erosion. Also, although there have been improvements in the developed world, soil loss has accelerated in the developing and newly industrialised nations.

11 UNCCD. Nations call for reversal of soil degradation, 28 January 2022. www.unccd.int/news-stories/stories/nations-call-reversal-soil-degradation.

12 World Resources Institute. Interactive map of eutrophication & hypoxia, 2013. https://bit.ly/3wp2nlT.

13 FAO. *The State of World Fisheries and Aquaculture 2020*. www.fao.org/3/ca9229en/online/ca9229en.html.

14 NOAA. Ocean acidification, 2021. https://bit.ly/3R5a9cM.

15 Breitberg D, Levin LA, Oschlies A, et al. Declining oxygen in the global ocean and coastal waters. *Science*, 2018, 359 (6371). https://doi.org/10.1126/science.aam7240.

16 IUCN. Marine plastic pollution, November 2021. www.iucn.org/resources/issues-brief/marine-plastic-pollution.

17 US Geological Survey. *Mineral Commodity Summaries 2021*. https://doi.org/10.3133/mcs2021.

18 UNEP International Resource Panel. *Decoupling Natural Resource use and Environmental Impacts from Economic Growth*, 2011. https://bit.ly/3QCYOAN.

19 Hayes A. Knowledge economy. *Investopedia*, 22 January 2021. www.investopedia.com/terms/k/knowledge-economy.asp.

20 Examples are the Ellen MacArthur Foundation, What is a circular economy?, https://bit.ly/3QOwmMv (accessed 24 August 2022) and Kate Raworth, Exploring doughnut economics, www.kateraworth.com/doughnut/ (accessed 24 August 2022).

21 Joita B. Recycling smartphones: Why do it, and how. *Techthelead*, 30 August 2021. https://techthelead.com/recycling-smartphones-why-do-it-and-how/.

22 McDonough W. Cradle to cradle, 2002. https://mcdonough.com/cradle-to-cradle/.

23 Wikipedia. Zero waste. https://en.wikipedia.org/wiki/Zero_waste (accessed 24 August 2022).

24 Cribb, *Food or War*.

25 Harvey F., Renewable energy has another record year of growth says IEA. *Guardian*, 1 December 2021. https://bit.ly/3R9Ed6P.

26 CASSE. Definition of steady state economy, 2022. https://steadystate.org/discover/definition-of-steady-state-economy/.

27 Pettinger T. Degrowth – definition, examples and criticisms. *Economics Help*, 26 April 2020. www.economicshelp.org/blog/164203/economics/degrowth/.

28 Schumacher EF. *Small Is Beautiful*. Harper Collins, 1973.

29 Global Footprint Network, 2022. www.footprintnetwork.org/.

30 Victor P & Rosenbluth G. Managing without growth. *Ecological Economics*, 2007, 61 (2–3), 492–504. https://doi.org/10.1016/j.ecolecon.2006.03.022.

31 Gleick P. The world water crisis and a path forward. In Council for the Human Future, *Delivering the Human Future*, March 2021, p. 11. https://bit.ly/3wX9p1N.

32   UN. Do you know all 17 SDGs?, 2022. https://sdgs.un.org/goals.
     See Goals 4, 13, 14, and 15.
33   Global Footprint Network. What is your ecological footprint?
     www.footprintcalculator.org (accessed 24 August 2022).
34   Marlin L. What is zero waste? How to slash the waste you produce.
     Greencoast, 19 May 2021. https://greencoast.org/what-is-zero-waste/.
35   Sustainable Table. What you can do. https://bit.ly/3wq3D8q
     (accessed 24 August 2022).

4   Nuclear Awakening

1   Spektor B. Russia declassifies footage of 'Tsar Bomba' – the most
    powerful nuclear bomb in history. *LiveScience*, 31 August 2020.
    https://bit.ly/3Alp3ow; also Reuters, Russia releases secret footage
    of 1961 Tsar Bomba hydrogen blast [online video clip],
    28 August 2020. https://youtu.be/YtCTzbh4mNQ.
2   Wikipedia. Historical nuclear weapons stockpiles and nuclear tests
    by country. https://bit.ly/3Ae6mD4 (accessed 24 August 2022).
3   See Federation of American Scientists. Status of world nuclear
    forces, 2022, https://fas.org/issues/nuclear-weapons/status-world-
    nuclear-forces/ and Ploughshares. World nuclear stockpile report,
    September 2021. https://ploughshares.org/world-nuclear-
    stockpile-report.
4   Bulletin of the Atomic Scientists, 27 January 2021. https://thebul
    letin.org/doomsday-clock/current-time/.
5   Ibid.
6   Kristensen HM & Korda M. China's nuclear missile silo
    expansion: From minimum deterrence to medium deterrence.
    Bulletin of the Atomic Scientists, 1 September 2021. https://bit.ly
    /3dXleyo.
7   Future of Life Institute. Accidental nuclear war: A timeline of close
    calls. https://futureoflife.org/background/nuclear-close-calls
    -a-timeline/ (accessed 24 August 2022).

8 Gorvett C. The nuclear mistakes that nearly caused World War Three. *BBC*, 10 August 2020. https://bbc.in/3dTlNco.

9 Toon A, Bardeen CG, Robock A, et al. Rapidly expanding nuclear arsenals in Pakistan and India portend regional and global catastrophe. *Science Advances*, 2019, 5 (10). http://doi.org/10.1126/sciadv.aay5478.

10 Cotta-Ramusino P. Regional rivalries and their implications for the security and nuclear non-proliferation regimes. In Maiani L, Jeanloz R, Lowenthal M, & Plastino W (eds.) *International Cooperation for Enhancing Nuclear Safety, Security, Safeguards and Non-proliferation*. Springer Proceedings in Physics, 2020, 243. Springer. https://doi.org/10.1007/978-3-030-42913-3_23.

11 UN Office for Disarmament Affairs. Treaty on the prohibition of nuclear weapons, 2017. www.un.org/disarmament/wmd/nuclear/tpnw/.

12 Campaign for Nuclear Disarmament (UK), 2022. https://cnduk.org/.

13 Greenpeace International. Nuclear weapons are illegal at last, 22 January 2021. https://bit.ly/3PKbYL1.

14 Rendle C. Why a world free of nuclear weapons is worth fighting for. International Committee of the Red Cross, 26 September 2021. www.icrc.org/en/document/world-free-from-nuclear-weapons.

15 The nine nuclear states are Russia, the United States, China, India, Pakistan, France, Britain, Israel, and North Korea.

16 Wikipedia. Nuclear umbrella, 2022. https://en.wikipedia.org/wiki/Nuclear_umbrella.

17 The White House. Joint statement of the leaders of the five nuclear-weapon states on preventing nuclear war and avoiding arms races, 3 January 2022. https://bit.ly/3cic9Q4.

18 Witmer S. Nuclear close calls. Nuclear Age Peace Foundation, 31 August 2017. www.wagingpeace.org/nuclear-close-calls/.

19 Schneider M. *The World Nuclear Industry Status Report 2018*. Mycle
    Schneider Consulting Project, 2018. www.worldnuclearreport.org
    /IMG/pdf/wnisr2018-v2-hr.pdf.

20 Arms Control Association. Chemical and biological weapons
    status at a glance, March 2022. www.armscontrol.org/factsheets/
    cbwprolif.

21 Wikipedia. Anti-nuclear movement, 2022. https://en
    .wikipedia.org/wiki/Anti-nuclear_movement.

22 Back from the Brink. Acting locally to prevent nuclear war:
    A back from the brink supporter toolkit, 2021. https://prevent
    nuclearwar.org/advocacy-tools/.

23 ICAN. Become a nuclear weapons ban advocate, 2021. https://
    icanw.org.au/action/.

24 Ploughshares. The real cost of nuclear weapons, 2022. www
    .ploughshares.org/real-cost-nuclear-weapons#prevent.

25 International Physicians for the Prevention of Nuclear War,
    2022. www.ippnw.org/.

26 Rendle, Why a world free of nuclear weapons is worth fighting
    for.

27 Wikipedia. Anti-nuclear organizations. https://en.wikipedia.org/
    wiki/Anti-nuclear_organizations (accessed 7 September 2022).

5   Cooling Earth

1 Beer T, Gill AM, & Moore PHR. Australian bushfire danger
   under changing climatic regimes. In Pearman GI (ed.)
   Greenhouse: Planning for Climate Change. CSIRO Australia, 1988, p.
   421.

2 Speth G. The Global 2000 report to the President. *Boston College
   Environmental Affairs Law Review*, 1980, 8 (4), 695–703. https://law
   digitalcommons.bc.edu/ealr/vol8/iss4/1/.

3 Intergovernmental Panel on Climate Change. *IPCC Sixth Assessment
   Report*, 9 August 2021. https://www.ipcc.ch/report/ar6/wg1/.

4 Moses A. 'Collapse of civilisation is the most likely outcome': Top climate scientists. *Voice of Action*, 8 June 2020. https://bit.ly/3qbQjkL.

5 Schellnhuber HJ. We are killing our best friends. Interview with Hans Joachim Schellnhuber, ZDF TV, Germany, 20 October 2019. https://bit.ly/3pExl5R.

6 Buis A. The atmosphere: Getting a handle on carbon dioxide. NASA JPL, 9 October 2019. https://go.nasa.gov/2IJJdyq.

7 Colins K. Climate tipping points, 6 September 2012. https://karinacollins.medium.com/climate-tipping-points-741693ce8a81.

8 Steffen W, Rockstrom J, Richardson K, et al. Trajectories of the earth system in the Anthropocene. *PNAS*, 2018, 115 (33), 8252–9. https://doi.org/10.1073/pnas.1810141115.

9 Raymond C, Matthews T, & Horton RM. The emergence of heat and humidity too severe for human tolerance. *Science Advances*, 2020, 6 (19). http://doi.org/10.1126/sciadv.aaw1838.

10 Canfield DE. A new model for Proterozoic ocean chemistry. *Nature*, 1998, 396, 450–3. https://doi.org/10.1038/24839.

11 Hoffman HJ. The Permian extinction – when life nearly came to an end. *National Geographic*, 6 June 2019. https://bit.ly/3AexQIK.

12 Ward PD. *Under a Green Sky*. Smithsonian/Collins, 2007.

13 Lombrana LM. In 2020, more people displaced by extreme climate than conflict. *Aljazeera*, 25 May 2021. https://bit.ly/3QLfeXO.

14 Council for the Human Future. *Delivering the Human Future*, March 2021, p. 12. https://bit.ly/3wX9p1N.

15 Desjardins J. This chart shows just how much energy the US is wasting. World Economic Forum, 25 May 25 2018. https://bit.ly/3QFXOvL.

16 In contrast with petroleum fuelled cars, which use only 20 per cent of the raw energy extracted to power them, electric vehicles use 90 per cent.

17 The World Counts. 1.3 billion tons of food is lost or wasted every year, 2022. https://bit.ly/3TLYEZV.

18 Intergovernmental Panel on Climate Change. *Climate Change 2022: Mitigation of Climate Change. Summary for Policymakers*, April 2022. https://bit.ly/3wnAgUg.

19 Monbiot G. Averting climate breakdown by restoring ecosystems: A call to action. Natural Climate Solutions, April 2019, www.naturalclimate.solutions/the-science; Wilson EO. *Half Earth: Our Planet's Fight for Life*. Liveright, 2016.

20 Blain L. Hydrogen 11 times worse than CO_2 for climate, says new report. *New Atlas*, 11 April 2022. https://newatlas.com/environ ment/hydrogen-greenhouse-gas/.

21 Ripple WJ, Wolf C, Newsome TM, et al. World scientists' warning of a climate emergency 2021. *BioScience*, 2021, 71 (9), 894–8. https://doi.org/10.1093/biosci/biab079.

22 See, for instance, City of Monash. How to reduce my carbon footprint, 4 August 2021. https://bit.ly/3qcgIic.

23 For example, Cho R. The 35 easiest ways to reduce your carbon footprint. Columbia Climate School, 27 December 2018. https://bit.ly/3dToR8q.

24 For example, Hilton C & Forest n. 50 Tips to cut down your carbon footprint. Global Giving, 17 May 2021. www .globalgiving.org/learn/reduce-carbon-footprint.

25 Mann ME. *The New Climate War – The Fight to Take Back Our Planet*. Public Affairs Books, 2021.

26 For example, Colson A. *Reduce Your Carbon Footprint: A Beginners Guide to Reducing Your Greenhouse Gas Emissions*. CreateSpace, 2015.

27 UNEP. 10 ways you can help fight the climate crisis, 4 May 2022. https://bit.ly/3dXS2Hy.

28 Project Drawdown. https://drawdown.org/ (accessed 24 August 2022).

29 Frischmann C & Chissell C. The powerful role of household actions in solving climate change. Project Drawdown, 27 October 2021. https://bit.ly/3CthZbT.

30 Global Footprint Network. What is your ecological footprint? www.footprintcalculator.org (accessed 24 August 2022).

31 Carbon Footprint. Carbon calculator. www.carbonfootprint.com/calculator.aspx (accessed 24 August 2022).

32 Kollmorgen A. Ethical investing guide. Choice, 17 June 2016. https://bit.ly/3CrX3SI.

33 Go Fossil Free. What is fossil fuel divestment?, 2021. https://gofossilfree.org/divestment/what-is-fossil-fuel-divestment/.

6   Clean Up the Planet

1 This is the upper end of the estimate I made in *Earth Detox*, based on the most reliable global sources I could find. It is probably a grave underestimate because no one has ever measured how much waste the world's mining industry generates and there is considerable uncertainty over issues such as soil erosion caused by agriculture, forestry, and development. Even the numbers for manufactured chemicals are unreliable, in view of a recent dramatic increase in the global inventory estimate and widespread industry secrecy.

2 Council for the Human Future. The megarisks, 2022 https://humanfuture.org/megarisks.

3 Wang Z, Walker GW, Muir DCG, & Nagatani-Yoshida K. Toward a global understanding of chemical pollution: a first comprehensive analysis of national and regional chemical inventories. *Environmental Science & Technology*, 2020, 54 (5), 2575–84. http://doi.org/10.1021/acs.est.9b06379.

4 UNEP. *Global Chemicals Outlook: Towards Sound Management of Chemicals*, 2013. https://bit.ly/3pIWEUr.

5 UNEP. *Global Chemical Outlook II: From Legacies to Innovative Solutions*, 2019. https://bit.ly/3CtO9nE.

6 WHO. Public health and environment, 2022. www.who.int/data/gho/data/themes/public-health-and-environment.

7 See Cribb, Earth Detox, pp. 20–1.

8   Harney E, Paterson S, Collin H, et al. Pollution induces epigenetic
     effects that are stably transmitted across multiple generations.
     Evolution Letters, 2022. https://doi.org/10.1002/evl3.273.
9   IUCN. Marine plastic pollution, November 2021. www.iucn.org
     /resources/issues-brief/marine-plastic-pollution.
10 Bratsburg B & Rogeburg O. Flynn effect and its reversal are both
     environmentally caused. *PNAS*, 2018, 115 (26) 6674–8. https://doi
     .org/10.1073/pnas.1718793115.
11 UNEP. Stockholm Convention: All POPs listed in the Stockholm
     Convention, 2022. www.pops.int/TheConvention/ThePOPs/
     AllPOPs/tabid/2509/Default.aspx.
12 WHO. Public health and environment, 2022.
13 Wikipedia. Hippocratic Oath, 2022. https://en.wikipedia.org
     /wiki/Hippocratic_Oath.
14 See, for example, the Environmental Working Group's *Shopper's
     Guide to Pesticides in Produce*, 2022. www.ewg.org/foodnews/sum
     mary.php.
15 Government of Victoria, Australia. Pesticides and other
     chemicals in food. https://bit.ly/3cmrUWe (accessed
     25 August 2022).
16 Volatile organic compounds (VOCs) are carbon-based chemicals
     in the form of gases and liquids. They are usually produced from
     fossil fuels and are often highly toxic.

7   Preventing Pandemics

1   Huang C, Wang Y, Li, X, et al. Clinical features of patients infected
     with 2019 novel coronavirus in Wuhan, China. *The Lancet*, 2020,
     395 (10223), 497–506.
     https://doi.org/10.1016/S0140-6736(20)30183-5.
2   WHO. 14.9 million excess deaths associated with the COVID-19
     pandemic in 2020 and 2021, 5 May 2022. https://bit.ly/3AkFjWD.
3   WHO. Data on the size of the HIV epidemic. https://bit.ly/
     3wWJl6Y (accessed 6 September 2022).

4   Institute of Medicine (US). *Microbial Evolution and Co-Adaptation: A Tribute to the Life and Scientific Legacies of Joshua Lederberg: Workshop Summary*. National Academies Press, 2009.

5   Harvard Global Health Institute. Preventing pandemics at the source. https://bit.ly/3ANcu6A (accessed 25 August 2022).

6   Ritchie H & Roser M. Meat and dairy production. Our World in Data, November 2019. https://ourworldindata.org/meat-production.

7   Madhav N, Oppenheim B, Gallivan M, et al. Pandemics: Risks, impacts, and mitigation. In Jamison DT, Gelband H, Horton S, et al. (eds.). *Disease Control Priorities*. The International Bank for Reconstruction and Development/World Bank, 2017, chapter 17.

8   An interesting aspect of this is the surge in infections that is observed after 'lockdowns' end, which researchers attribute in part to an increase in promiscuity. See Neri I & Gammaitoni L. Role of fluctuations in epidemic resurgence after a lockdown. *Scientific Reports*, 2021, 11, 6452. https://doi.org/10.1038/s41598-021-85808-z.

9   Wang Q, Han J, Chang H, Wang C, & Lichtfouse E. Society organization, not pathogenic viruses, is the fundamental cause of pandemics. *Environmental Chemistry Letters*, 2022, 20, 1545–51. https://doi.org/10.1007/s10311-021-01346-0.

10 Kossaify E. WHO official: 'Next pandemic may be more severe'. *Arab News*, 29 December 2020. www.arabnews.com/node/1784546/world.

11 WHO. World Health Assembly agrees to launch process to develop historic global accord on pandemic prevention, preparedness and response, 1 December 2021. https://bit.ly/3R6t5Yy.

12 Ferro S. The anti-spitting campaigns designed to stop the spread of tuberculosis. *Mental Floss*, 13 November 2018. https://bit.ly/3dWatvW.

13 Deigin Y. Thunder out of China. *Inference*, February 2022. https://inference-review.com/article/thunder-out-of-china.

14   Burki T. Ban on gain-of-function studies ends. *The Lancet*, 2018, 18 (2), 148–9. https://doi.org/10.1016/S1473-3099(18)30006-9.

15   Dyer O. Covid-19: Unvaccinated face 11 times risk of death from delta variant, CDC data show. *British Medical Journal*, 2021, 374 (2282). https://doi.org/10.1136/bmj.n2282.

16   WHO. Vaccine-preventable diseases (including pipeline vaccines). www.who.int/teams/immunization-vaccines-and-biologicals/diseases (accessed 25 August 2022).

17   WHO. Managing the Covid Infodemic: Promoting healthy behaviours and mitigating the harm from misinformation and disinformation, 23 September 2020. https://bit.ly/3dPbki1.

8   Renewable Food

1   Cribb, *Food or War*.

2   Leisner CP. Review: Climate change impacts on food security – focus on perennial cropping systems and nutritional value. *Plant Science*, 2020, 293, 110412. https://doi.org/10.1016/j.plantsci.2020.110412.

3   Aquaculture is the culture of plants and fish in water for food and other uses. Aquaponics is the growing of fish and plants on floating rafts in the same tank.
     Agritecture is a new discipline that combines building design (architecture) with the production of plants for food and other purposes. Cellular agriculture is the production of food from plant or animal cell cultures, grown under controlled conditions in a large container.

4   Food Plant Solutions. Bruce French. https://foodplantsolutions.org/bruce-french/ (accessed 26 August 2022).

5   Ratner BD, Meinzen-Dick R, Hellin J, et al. Addressing conflict through collective action in natural resource management: A synthesis of experience. CAPRi Working Paper No. 112, International Food Policy Research Institute (IFPRI), August 2013. https://bit.ly/3QXGZwr.

6   Cribb, *Food or War*.

9 'One Child Fewer'

1 Rozell N and Chay D. St. Matthew Island – overshoot & collapse. *Resilience*, November 2003. https://bit.ly/3dVNv8j.

2 Røsjø B. The growth and decline in Easter Island's population is a lesson for our future. University of Oslo, 30 September 2020. https://bit.ly/3Asd4p2. Over the last 800 years, the population of Rapa Nui underwent a series of crises. The study shows these crises were linked to the long-term effects of climate change on the island's ability to produce food.

3 Rushkoff D. The super-rich 'preppers' planning to save themselves from the apocalypse. *Guardian*, 4 September 2022. https://bit.ly /3eyGYAY.

4 Biology Dictionary. Overpopulation, 4 October 2019. https://bio logydictionary.net/overpopulation/.

5 Worldometer. World population projections. www.world ometers.info/world-population/world-population-projections/ #:~:text=World%20Population%20Projections%20%20%20% 20Year%20,%20%2054%20%2030%20more%20rows%20 (accessed 26 August 2022).

6 UNHCR. Migrants and refugees, 29 December 2021. https://news .un.org/en/story/2021/12/1108472.

7 Edmond C. Global migration, by the numbers: Who migrates, where they go and why. World Economic Forum, 10 January 2020. https://bit.ly/3Kr2frV.

8 World Bank. Climate change could force 216 million people to migrate within their own countries by 2050, 13 September 2021. https://bit.ly/3Tjwwgh.

9 Bassetti F. Environmental migrants: Up to 1 billion by 2050. *Foresight*, 22 May 2019. https://bit.ly/3pIpdS1.

10 Kanem N, The economic benefits of family planning. World Economic Forum, 26 July 2018. https://bit.ly/3PS7MZC.

11 O'Sullivan J. Facing up to overpopulation, March 2021. Online video clip. https://youtu.be/7VJSWI5mPPI.

12 Ibid.

13 Ibid.

14 UN Population Fund. Family planning. www.unfpa.org/family-planning (accessed 26 August 2022).

15 O'Sullivan, Facing up to overpopulation.

16 Wynes S & Nicholas KA. The climate mitigation gap: Education and government recommendations miss the most effective individual actions. *Environmental Research Letters*, 2017, 12 (7), 074024. https://iopscience.iop.org/article/10.1088/1748-9326/aa7541.

17 Cain S. Why a generation is choosing to be child-free. *Guardian*, 25 July 2020. https://bit.ly/3pJSLPd.

18 Neal JW & Neal ZP. Prevalence and characteristics of childfree adults in Michigan (USA). *PLoS One*, 2021, 16 (6), e0252528. https://doi.org/10.1371/journal.pone.0252528.

10   Healing Technological Mayhem

1 Bostrom N. The vulnerable world hypothesis. *Global Policy*, 2019, 10 (4), 1. https://doi.org/10.1111/1758-5899.12718.

2 Ibid.

3 Green BP. Emerging technologies, catastrophic risks, and ethics: Three strategies for reducing risk. *IEEE International Symposium on Ethics in Engineering, Science and Technology (ETHICS)*, 2016, 1–7. https://doi.org/10.1109/ETHICS.2016.7560046.

4 DiEuliis D, Ellington AD, Gronvall GK, & Imperiale MJ. Does biotechnology pose new catastrophic risks? *Current Topics in Microbiology and Immunology*, 2019, 424, 107–19. https://doi.org/10.1007/82_2019_177.

5 Synthetic biology. www.sciencedirect.com/topics/engineering/synthetic-biology (accessed 7 September 2022).

6 FBI. Amerithrax or anthrax investigation, 2011. www.fbi.gov/history/famous-cases/amerithrax-or-anthrax-investigation.

7 Sample I. Scientists condemn 'crazy, dangerous' creation of deadly airborne flu virus. *Guardian*, 12 June 2014. https://bit.ly/3Q4XDsT.

8   NBC News. World Health Organization chief says it was 'premature' to rule out Covid lab leak. Associated Press, 16 June 2021. https://nbcnews.to/3dYKKmC.

9   Tang S, Wang M, Germ KE, et al. Health implications of engineered nanoparticles in infants and children. *World Journal of Pediatrics*, 2015, 11, 197–206. https://doi.org/10.1007/s12519-015-0028-0.

10 Center for Responsible Nanotechnology. Dangers of molecular manufacturing. http://crnano.org/dangers.htm (accessed 31 August 2022).

11 Molecular self-assembly. www.sciencedirect.com/topics/chemistry/molecular-self-assembly (accessed 7 September 2022).

12 Cellan-Jones R. Stephen Hawking warns artificial intelligence could end mankind. *BBC News*, 2 December 2014. www.bbc.com/news/technology-30290540.

13 Thomas M. 7 dangerous risks of artificial intelligence. *Built In*, 6 July2021. https://builtin.com/artificial-intelligence/risks-of-artificial-intelligence.

14 Liptak B. Can we control artificial intelligence? *Control*, 27 January 2020. www.controlglobal.com/articles/2020/can-we-control-artificial-intelligence/.

15 Sychev V. The threat of killer robots. UNESCO, 2018. https://en.unesco.org/courier/2018-3/threat-killer-robots.

16 Schue P. Robot armies and the future of warfare. *Clio's Nightmare*, 26 June 2021. https://bit.ly/3ADI7hK.

17 Dawes J. Humanity's final arms race: UN fails to agree on 'killer robot' ban. *Common Dreams*, 30 December 2021. https://bit.ly/3RkEi83.

18 Human Rights Watch. Killer robots: Military powers stymie ban, 19 December 2021. www.hrw.org/news/2021/12/19/killer-robots-military-powers-stymie-ban.

19 Glover D. UK climate activists jailed, with controversial protest restrictions coming. *Courthouse News Service*, 19 November 2021. https://bit.ly/3wEOr7p.

20   Cribb JHJ. The end of freedom. Surviving C21 (blog), 1 February
     2018. https://juliancribb.blog/2018/02/01/the-end-of-freedom/.

21   Estes R. Large hadron collider: What's the risk? *On-Screen Scientist*,
     8 September 2008. https://onscreen-scientist.com/?p=34.

22   World Economic Forum. *The Global Risks Report 2022*. www
     .weforum.org/reports/global-risks-report-2022.

23   UN Oceans & Law of the Sea. United Nations Convention on the
     Law of the Sea of 10 December 1982. https://bit.ly/3qh8o0R.

24   Dade C. Reestablishing the Congressional Office of Technology
     Assessment. *Journal of Science Policy & Governance*, 2019, 15 (1).
     www.sciencepolicyjournal.org/uploads/5/4/3/4/5434385/
     dade_jspg_v15.pdf.

25   IPCC. www.ipcc.ch/about/ (accessed 7 September 2022).

26   Wilson G. Minimizing global catastrophic and existential risks
     from emerging technologies through international law. *Virginia
     Environmental Law Journal*. 2013, 31 (2), 307–36. www.jstor.org/
     stable/44679544.

27   Wikipedia. *Primum Non Nocere*. https://en.wikipedia.org/wiki/
     Primum_non_nocere (accessed 31 August 2022).

28   See Cribb, *Earth Detox*, pp. 240–2.

29   UN. Universal Declaration of Human Rights, 10 December 1948.
     www.un.org/en/about-us/universal-declaration-of-human-
     rights.

11   Ending the Age of Deceit

1   Doomsday Clock, 2021. https://thebulletin.org/doomsday-clock
    /current-time/.

2   West J & Bergstrom C. Misinformation in and about science.
    Proceedings of the National Academy of Sciences, 2021, 118 (15),
    e1912444117. https://doi.org/10.1073/pnas.1912444117.

3   Goldman R. Reading fake news, Pakistani minister directs nuclear
    threat at Israel. *New York Times*, 24 December 2016. https://nyti.ms
    /3B1d3do.

4 US Senate. Report of the Select Committee on Intelligence, United States Senate on Russian active measures campaigns and interference in the 2016 U.S. election. Senate Report 116–290, US Government, 2019.

5 Van der Linden S, Leiserowitz A, Rosenthal S, & Maibach E. Inoculating the public against misinformation about climate change. *Global Challenges*, 2017, 1, 1600008.

6 Salmon DA, Dudley MZ, Glanz JM, & Omer SB. Vaccine hesitancy: Causes, consequences, and a call to action. *Vaccine*, 2015, 33, D66–D71.

7 Editorial. The Covid-19 infodemic. *The Lancet*, 2020, 20 (8), 875. https://doi.org/10.1016/S1473-3099(20)30565-X.

8 West & Bergstrom, Misinformation in and about science.

9 Johnson AG, Amin AB, Ali AR, et al. *Covid-19 Incidence and Death Rates Among Unvaccinated and Fully Vaccinated Adults with and without Booster Doses during Periods of Delta and Omicron Variant Emergence – 25 U.S. Jurisdictions, April 4–December 25, 2021*. Centers for Disease Control and Prevention, 28 January 2022. www.cdc.gov/mmwr/volumes/71/wr/mm7104e2.htm.

10 Lin H. The existential threat from cyber-enabled information warfare. *Bulletin of the Atomic Scientists*, 2019, 75 (4), 187–96. https://doi.org/10.1080/00963402.2019.1629574.

11 Novella S. The misinformation dilemma. *Science-Based Medicine*, 19 January 2022. https://sciencebasedmedicine.org/the-misinformation-dilemma/.

12 Laville S. Top oil firms spending millions lobbying to block climate change policies, says report. *Guardian*, 22 March 2019. https://bit.ly/3ee0Vwp.

13 Cook J, Supran G, Lewandowsky S, Oreskes N, & Maibach E. *America Misled: How the fossil fuel industry deliberately misled Americans about climate change*. Center for Climate Change Communication, George Mason University, 2019. www.climatechangecommunication.org/america-misled/.

14 Oreskes N & Conway EM. *Merchants of Doubt: How a Handful of Scientists Obscured the Truth on Issues from Tobacco Smoke to Global Warming.* Bloomsbury, 2018.

15 The countries are: Angola, Argentina, Armenia, Australia, Austria, Azerbaijan, Bahrain, Brazil, Cambodia, China, Colombia, Cuba, Czech Republic, Ecuador, Egypt, Germany, Hungary, India, Iran, Israel, Italy, Kenya, Kyrgyzstan, Malaysia, Mexico, Myanmar, Netherlands, Nigeria, North Korea, Pakistan, Philippines, Poland, Russia, Saudi Arabia, Serbia, South Africa, South Korea, Syria, Taiwan, Thailand, Turkey, UK, Ukraine, United Arab Emirates, United States, Venezuela, Vietnam, and Zimbabwe.

16 Bradshaw S & Howard PN. *Challenging Truth and Trust: A Global Inventory of Organized Social Media Manipulation.* Oxford Internet Institute, Oxford University, 2018. https://bit.ly/3RsW60Q.

17 Hotez P. The antiscience movement is escalating, going global and killing thousands. *Scientific American*, 29 March 2021. https://bit.ly/3cAejea.

18 Investigate Europe. The disinformation machine, April 2019. https://bit.ly/3ACbgtW.

19 Bump P. The unique role of Fox News in the misinformation universe. *Washington Post*, 8 November 2021. https://wapo.st/3egZ6Pi.

20 Vanian J. Russian disinformation campaigns are trying to sow distrust of Covid vaccines, study finds. *Fortune*, 23 July 2021. https://bit.ly/3Q3fSPf.

21 Twaij A. The Republican Party of misinformation. *Newsweek*, 3 January 2021. www.newsweek.com/republican-party-misinformation-opinion-1572364.

22 Bratsburg B & Rogeburg O. Flynn effect and its reversal are both environmentally caused. *PNAS*, 2018, 115 (26), 6674–8. https://doi.org/10.1073/pnas.171879311.

23 Buchanan T. How to reduce the spread of fake news – by doing nothing. *The Conversation*, 19 December 2020. https://bit.ly/3wOkORr.

12 Who Are We, Really?

1 Thompson A. Oldest known jellyfish fossils found. *LiveScience*, 31 October 2007. https://bit.ly/3cFfQj0.

2 As an anatomist and taxonomist, Linnaeus was unable to discern any physical features that separated humans from other apes, and so chose 'self-awareness' as the defining characteristic of humanity. He expressed this with the Latin word *sapiens*, which more commonly means 'wise', and this has become the modern translation. Our species name may therefore be a linguistic accident.

3 Cribb J. New name needed for unwise Homo? *Nature*, 2011, 476, 282. https://doi.org/10.1038/476282b.

4 Cribb, *Surviving the 21st Century*, pp. 210–11.

5 Cribb J. Why *Homo sapiens* needs a new name. Surviving C21 (blog), 12 December 2016. https://bit.ly/3KRNNcZ.

13 An Earth Standard Currency

1 Global Footprint Network, 2021. www.footprintnetwork.org/.

2 Stockholm Resilience Centre. The nine global boundaries, 2021. https://bit.ly/3TJLWee.

3 World Population Review. Legatum Global Prosperity Index, 2021. https://worldpopulationreview.com/country-rankings/legatum-prosperity-index.

4 Wellbeing Economy Alliance. Bhutan – Gross National Happiness Index, 2021. https://weall.org/resource/bhutan-gross-national-happiness-index.

5 Threatened Species Index, 2021. https://tsx.org.au/.

6 Minderoo Foundation, 2021. Global Fishing Index. www .minderoo.org/global-fishing-index/#key-findings.

7 Cardwell MR. Attenborough: Poorer countries are just as concerned about the environment. *Guardian*, 16 October 2013. https://bit.ly/3q3GZiM.

14 Tools for Repairing the Earth

1 See, for example, Magalhaes et al., *Safe Operating Space Treaty*.
2 UN. Transforming our world: The 2030 Agenda for Sustainable Development, 2015. https://sdgs.un.org/2030agenda.
3 UN. Sustainable development goals: The 17 Goals, 2022. https://sdgs.un.org/goals.
4 UN. Universal Declaration of Human Rights, 1948. www.un.org/en/about-us/universal-declaration-of-human-rights.
5 Roy EA. New Zealand river granted same legal rights as human being. *Guardian*, 16 March 2017. https://bit.ly/3AAyXmg.
6 Truth and Reconciliation Commission, South Africa, 2022. www.justice.gov.za/trc/.

15 Think Like a Human, Act Like a Species

1 World Internet Statistics. Internet usage statistics: The internet big picture, 2022. https://internetworldstats.com/stats.htm.
2 Jones BO. *What Is to Be Done?* Scribe Press, 2021.
3 UN. Sustainable development goals, 2022. https://sdgs.un.org/goals.
4 Lovelock JE. *The Revenge of Gaia*. Penguin, 2007.
5 Cribb JHJ. Humanity: Sailing into a stagnant ocean. Surviving C21 (blog), 25 February 2022. https://bit.ly/3RiGlKk.

INDEX